Cosmologies of the Anthropocene

This book engages with the classic philosophical question of mind and matter, seeking to show its altered meaning and acuteness in the era of the Anthropocene. Arguing that matter, and, more broadly, the natural world, has been misconceived since Descartes, it explores the devastating impact that this has had in practice in the West. As such, alternatives are needed, whether philosophical ones such as those offered by figures such as Whitehead and Nagel, or posthumanist ones such as those developed by Barad and Latour. Drawing on recent anthropological work ignored by philosophers and sociologists alike, the author considers a radical alternative cosmology: animism understood as panpsychism in practice. This understanding of mind and matter, of culture and nature, is then turned against present-day posthumanist critiques of what the Anthropocene amounts to, showing them up as philosophically misguided, politically mute, and ethically wanting. A ground-breaking reconceptualization of the natural world and our treatment of it, *Cosmologies of the Anthropocene* will appeal to scholars of sociology, social theory, philosophy and anthropology with interests in our understanding of and relationship with nature.

Arne Johan Vetlesen is Professor of Philosophy at the University of Oslo, Norway. He is the author of *A Philosophy of Pain; Evil and Human Agency: Understanding Collective Evildoing; Perception, Empathy, and Judgment*; and *The Denial of Nature: Environmental Philosophy in the Era of Global Capitalism.*

Morality, Society and Culture

The Morality, Society and Culture series publishes rigorous scholarly work exploring how moral questioning and action have been transformed in contemporary social relationships and by contemporary culture. Can cultural texts such as films, television broadcasts and art be vehicles for moral demands? Do we learn what it means to be 'good' from soap opera and advertising? If cultural texts are forms of moral mimesis, then are the standards of the 'right' and 'good' dependent on external considerations of cultural visibility and social relevance – and if so, how are some moral issues made visible or invisible, relevant or irrelevant?

Now that morality has become cultural and is amenable to sociological and cultural study as well as philosophical investigation, this series explores how and to what effect moral questioning, action and debate are inextricably entwined with contemporary social and cultural forms, texts and institutions. The books in this series offer new understandings of the connection of morality, society and culture, analyse key contemporary events, and establish new methodologies.

Series Editor

Keith Tester is Professor at The *Thesis Eleven* Centre for Cultural Sociology at LaTrobe University, Australia and Senior Fellow at the Centrum Myśli Jana Pawła II, Warsaw, Poland. He is the author of *Humanitarianism and Modern Culture*, *Panic*, *Eric Rohmer: Film as Theology*, *Moral Culture*, *The Social Thought of Zygmunt Bauman*, and *Animals and Society*, co-author of *Conversations with Zygmunt Bauman* and *What Use is Sociology?* and co-editor of *Utopia: Social Theory and the Future*.

Titles in this series

Everyday Moralities
Doing it Ourselves in an Age of Uncertainty
Nicholas Hookway

Cosmologies of the Anthropocene
Panpsychism, Animism, and the Limits of Posthumanism
Arne Johan Vetlesen

For more information about this series, please visit: www.routledge.com/Morality-Society-and-Culture/book-series/ASHSER1429

Cosmologies of the Anthropocene

Panpsychism, Animism, and the Limits of Posthumanism

Arne Johan Vetlesen

Routledge
Taylor & Francis Group
LONDON AND NEW YORK

First published 2019
by Routledge
2 Park Square, Milton Park, Abingdon, Oxon OX14 4RN

and by Routledge
52 Vanderbilt Avenue, New York, NY 10017

Routledge is an imprint of the Taylor & Francis Group, an informa business

British Library Cataloguing-in-Publication Data
A catalogue record for this book is available from the British Library

Library of Congress Cataloging-in-Publication Data
Names: Vetlesen, Arne Johan, 1960– author.
Title: Cosmologies of the Anthropocene : panpsychism, animism, and the limits of posthumanism / Arne Johan Vetlesen.
Description: 1 [edition]. | New York : Routledge, 2019. | Series: Morality, society, and culture | Includes bibliographical references and index.
Identifiers: LCCN 2018055567 (print) | LCCN 2019005640 (ebook) | ISBN 9780429060564 (ebk) | ISBN 9780429595387 (web pdf) | ISBN 9780429594090 (epub) | ISBN 9780429592805 (mobi/kindle) | ISBN 9780367182922 (hbk)
Subjects: LCSH: Cosmology. | Panpsychism. | Realism. | Humanism.
Classification: LCC BD511 (ebook) | LCC BD511 .V48 2019 (print) | DDC 113—dc23
LC record available at https://lccn.loc.gov/2018055567

ISBN: 978-0-367-18292-2 (hbk)
ISBN: 978-0-429-06056-4 (ebk)

Typeset in Times New Roman
by Apex CoVantage, LLC

Contents

Preface

The sequel to my *The Denial of Nature*, this book was conceived during a course I gave on metaphysics, with special emphasis on panpsychism, at the University of Oslo. The topic was new to me, but the keen interest taken by the students in discussing it shaped (and hopefully sharpened) my thinking and inspired me to draft what was to become the first three chapters of the present work.

As the opening chapter makes clear, I have in recent years become increasingly dissatisfied which much of what passes as academic philosophy these days. My chosen discipline appears, with some notable exceptions, "*weltfremd*", having retreated from the messy world of public affairs to the purity and closed circuit of publish (in journals read by a handful of colleagues) or perish – thereby rendering itself irrelevant to the outside world and in effect digging its own grave as a worthwhile undertaking of more than academic importance.

This development is particularly frustrating given the advent of the Anthropocene and the host of problems (not challenges and new opportunities) that go with it: global warming, the extinction of species, and the loss of biological and cultural diversity.

I have for many years been convinced that the two losses, the biological and the cultural, are interconnected, inextricably and substantially so. But where to turn to account for this largely neglected fact? It is overlooked in disciplines that are too specialized to be aware of more than the one dimension or object domain delegated to them, reproducing within the academy – as its division of labour – the dualism between mind and matter, humans and (other) animals, culture and nature, launched by Descartes all those years ago – denounced over and over again in theory, to be sure, but *retained in practice*, as the Anthropocene reminds us, frighteningly late in the day at that.

Again, where to turn not only to account for this sorry state of affairs, but to correct and, if possible, overcome it? To cosmology, I suggest.

At what proved a decisive moment, my longtime colleague Thomas Hylland Eriksen suggested I read the British anthropologist Tim Ingold. I did, picking his until then unopened *The Perception of the Environment* from my shelf. Thus was born the thesis that informs the second half of this work: that animism is panpsychism in practice; and that this helps illuminate what the Anthropocene is all about.

I am grateful to Thomas for his comments on the synthesis of philosophy and anthropology that I attempt here. Deepfelt thanks also go to Per Bjørn Foros and Monica Bjermeland; both gave thorough feedback to the drafts I sent them, emboldening me when I entered into unknown territory and was about to lose my nerve. I also extend my thanks to the colleagues, students, and friends (in various combinations) who have helped shape my thinking on the topics discussed here: Trond Arnesen, Elling Ulvestad, Espen Hammer, Alastair Hannay, Petter Thorvaldsen, Carlos Joly, Martin Lee Mueller, Ingerid Straume, Per Nortvedt, Rasmus Willig, Dag Olav Hessen, Odin Lysaker, and Sigurd Hverven. Keith Tester, with his characteristic combination of kindness, enthusiasm, and critical scrutiny read the whole manuscript and promptly directed it to the right person at the right publisher: Neil Jordan at Routledge. In a more general fashion to do with my overall philosophical outlook, I have benefited from conversations with Charles Taylor, Konrad Ott, and Hartmut Rosa (who kindly invited me to a conference on "New Materialism" in Erfurt in November 2017), and – last but not least – from disagreements old and new with my teacher Jürgen Habermas.

Horten, Norway

Introduction

From anthropocentrism to the Anthropocene

In his book *Panpsychism in the West*, David Skrbina writes:

> The mechanistic worldview is deeply embedded in our collective psyche. For several hundred years the dominant orthodoxy has implicitly assumed that inanimate things are fundamentally devoid of mental qualities. This view has become integrated into our science, our literature, and our arts. Ultimately it has incorporated itself into our deepest social values, and thus become reflected in our collective actions. We treat nature as an impersonal thing or collection of things, without spontaneity, without intrinsic value, without "rights" of any kind. Natural resources, plant and animal species have been exploited for maximal short-term human benefit. Such mindless entities are seen as deserving of no particular respect or moral consideration. They exist to be collected, manipulated, dissected, and remade.
>
> (Skrbina 2009: 265)

In quoting this portrait of the mechanistic worldview, I am well aware that its presuppositions – say, in the form of the Cartesian mind/matter dualism – have been subject to massive criticism ever since its breakthrough in the sixteenth century. And yet, despite attempts within both philosophy and the natural sciences to show the notion of nature as "an impersonal thing or collection of facts" to be simply untrue to the facts, as a society, we continue to exploit natural resources, plants, and animal species with impunity; far from bringing the exploitation of the natural world to a halt, we are busy speeding it up. By restricting the capacities for mind and soul, intelligence and reason, spontaneity and purpose to human beings, the mechanistic worldview has helped entrench anthropocentrism – literally, human-centeredness – in all key domains of modern Western society to this very day.

Anthropocentrism is not merely, or even primarily, descriptive. It is normative in postulating that human beings are superior to all other beings and forms of life on Earth, thus meriting a moral standing denied everything nonhuman. This normativity informs anthropocentrism as a *practice*, as acted upon individually and collectively. The practices to which anthropocentrism gives rise, and which it helps legitimize, span the entire range of institutions characterizing modern society – initially, Western society, today the whole world. Whether it be the institutions of economy, of politics, of education, of health or of law, they are either exclusively or primarily preoccupied with *human* agents and their perceived interests and needs. That this is so, and ought to remain so, serves as the pivotal "reality principle" on which the socialization of every new generation is premised: in the course of childhood the anthropocentric point of view is internalized so as to become second nature – always presupposed in relating to other-than-humans as well as fellow humans, never seriously questioned by adults taken seriously by others. Internalized from early childhood is a sense of being profoundly different from all other living beings, promoting an ethos of entitlement in treating all such beings as mere means for human ends, as so many "resources" whose alleged indifference to what we do to (against) them conveniently justifies our own, institutionally as well as individually. Owing to its seminal role in guiding all our practices, anthropocentrism is one of the most deep-seated and pervasive features of modern culture and of ourselves as products and reproducers of that culture.

Future historians may well find that the most remarkable thing about this culture is how profoundly it came to change its declared "other", nature. Paradoxically and unexpectedly, anthropocentrism as acted upon and helping institutionalize the entire set of practices devoted to humankind's exploitation of nature has had the effect of giving the lie to the notion from which it started out: that culture and nature, in a timeless, static, and once-and-for-all manner, constitute two perfectly separate domains of reality, each internally homogeneous and incommensurable with the other; whereby culture connotes subjectivity, activity, signification, meaning, identity, and purpose, and nature connotes object, passivity, body, instinct, animal, and law-like. As every student of the social sciences knows, Max Weber put it unequivocally: "Culture is a finite segment of the meaningless infinity of the world process, a segment on which *human beings* confer meaning and significance" (1949: 81). Weber is echoed in anthropologist Clifford Geertz's oft-cited formulation that "man is an animal suspended in webs of significance he himself has spun" (1973: 5), inspiring Fredrik Barth's assertion that in dealing with a (any) group's cosmology, "we are dealing with meaning which is *conferred* on a sector of the world" (1987: 69), a sector held to be void of meaning prior to and independently of humans actively conferring it. Importantly, the alleged matters of fact hardly conceal a highly charged normativity: everything to do with *agency* and full-fledged moral standing is reserved to the humans constituting culture, and everything to do with being acted on, understood as passivity, as being at the receiving end, to its other, nature.

Why do I claim that this view about culture and nature forming two independent domains has been given the lie? As the chapters that follow will make clear, the simple yet essential division between a realm that exhibits agency and another – simply designated "nature" despite its enormous heterogeneity – that is merely passive, itself lacking in agency, does not hold water. That it does not, however, is not to be considered a timeless fact. Rather, the issue of the agency of – in – the natural world is a thoroughly historical one, presently forcing us to confront the spectrum of other-than-human agencies in an altogether different and more direct manner than ever in (human) history, agencies of a kind – as it now turns out – that are both potent and damaged, depending on the ways in which *our* activities impact on them.

If this sounds lofty and abstract, it isn't. I am getting at a simple fact, albeit a fairly recent and often misunderstood one, partly since the intellectual tradition that we still carry makes us singularly ill-prepared to recognize it, and partly for not-so-subtle psychological reasons, denial looming large among them. The fact is that as acted upon over the last few centuries, anthropocentrism has paved the way for the entry into the Anthropocene. If anthropocentrism is the philosophy, or – writ large – cosmology, now gone global, the Anthropocene signifies the new, and anything but fake, facts on the ground brought about in its course. The connection between the two is an intimate one, yet rarely stated as such, as a matter of cause and effect.

As is well known, the term *Anthropocene* was coined as recently as 2002 by the Dutch chemist Paul Crutzen. In a short paper, "Geology of Mankind", published in *Nature*, Crutzen wrote: "It seems appropriate to assign the term 'Anthropocene' to the present, in many ways human-dominated, geological epoch" (Crutzen 2002: 23). The shift from the geological epoch called the Holocene to the Anthropocene is due to the fact that the influence of humanity now outweighs the impact of the most powerful natural forces on the functioning of the Earth system. Among the many geologic-scale changes we humans have effected, Crutzen cites the following:

- Human activity has transformed between a third and a half of the land surface of the planet.
- Most of the world's major rivers have been dammed or diverted.
- Fisheries remove more than a third of the primary production of the ocean's coastal waters.
- Humans use more than half of the world's readily accessible freshwater runoff.

Since Crutzen's list was drawn up in 2002, let me add some recent findings:

- Earth Overshoot Day 2018: August 1. Humanity exhausts Earth's budget in seven months, demanding more renewable resources than the planet can provide for an entire year. We over-utilize and over-exploit nature's riches – resources, stocks – to such an extent that they are increasingly unable to regenerate. Human-caused overshoot produces degradation, depletion, and extinction.
- As reported by WWF, the number of wild animals on Earth has halved in the last forty years. Creatures across land, rivers, and the seas are being decimated as humans kill them for food in unsustainable numbers, while polluting, fragmenting, and destroying their habitats.

It seems fitting that scholars from the natural sciences are at the forefront of these developments, being the first to observe the shift from the Holocene to the Anthropocene in its empirical reality, and therefore better placed than other scientists to predict that as a result of anthropogenic factors the global climate will "depart significantly from natural behavior for many millennia to come", to quote Crutzen's paper. A comprehensive study led by Professor of Ecology Gerardo Ceballos, using a sample of 27 600 terrestrial vertebrate species, and a more detailed analysis of 177 mammal species, published in *Proceedings of the National Academy of Sciences* in July 2017, concludes on a note of alarm that breaks with the "value neutrality" associated with the natural sciences: "The biological annihilation we now observe will without doubt have severe ecological, economic, and social consequences. Humanity will eventually pay an exceptionally high price for the ongoing decimation of the only habitat we know in the universe. The situation is so critical that it would be unethical not to use strong words" (Ceballos et al. 2017: 3; Carrington 2017: 12). The scientists continue:

> Dwindling population sizes and range shrinkages amount to a massive anthropogenic erosion of biodiversity and of the ecosystem services essential to civilization. [. . .] The ultimate drivers of the immediate causes of biotic destruction [are] human overpopulation and overconsumption, especially by the rich. These drivers, all of which trace to the fiction that perpetual growth can occur on a finite planet, are themselves increasing rapidly. All signs point to ever more powerful assaults on biodiversity in the next two decades, painting a dismal picture of the future of life, including human life.
>
> (Ceballos et al. 2017: 7, 8)

Ceballos and his co-authors alert us what is truly novel about the Anthropocene, namely what is presently becoming manifest – as so many observations and data – as the effects of human activities on the Earth, with enormously increased momentum since 1945 (known as the "Great Acceleration"; see Angus 2016). To be sure, simply tracing the Anthropocene to its origins in the anthropocentrism that pervades the mechanistic worldview paints much too general a picture: what the world has never seen before cannot be explained in terms of the long familiar, and so as more of the same, if only in a qualitatively novel fashion. For this reason, attempts to locate the onset of what is manifest today as the systemic imbalance between humans and (the rest of) nature in, say, the domestication of plants and animals, in the emergence of agriculture, or in the class-based conflict between capital and labour (Marx), as articulated in Jason Moore's (2015) notion of the "Capitalocene", all miss out on the novelty in question. That said, I hasten to add that there is indeed a strong causal connection between present-day global capitalism and the ecological destruction at the heart of the Anthropocene, a connection clearly implied in the explanation given by Ceballos et al above; a connection I shall, however, for the most part leave aside in this work, having explored it at length in *The Denial of Nature: Environmental Philosophy in the Era of Global Capitalism*. Likewise, the current attempt from the other side of the spectrum, the so-called ecomodernists, to present the Anthropocene as something good, as an opportunity to fully assert human mastery on planet Earth and celebrate the dawn of a new generation of sciences, such as geoengineering, also distracts us from what is happening and its true implications.

Far from being an old story poured into new bottles and given a sexy label, then, the Anthropocene is an instance of *catastrophism*, as pointed out by Clive Hamilton (2017: 26). Catastrophism captures the first thing to say about it, never to be lost from sight. It is, fundamentally, an unprecedented geological event, one lacking not only in predecessors but in valid comparisons and analogues as well, leaving us at a loss as to where to turn for lessons about how to understand it and – somehow – cope with it. It is an event that has precisely *not* occurred – even if starting slowly and locally – "ever since" this or that paradigm, in thought and in economic and technological practices, became triumphant centuries ago. As Hamilton rightly insists, the Anthropocene is a *rupture*. It is so in a geological and geophysical sense not to be conflated with a social, economic, or historical one, even though of course now heavily impacting all of them. And if we want to insist that geological history and human history have "converged", it does not mean that geology has become social science (as Moore's Marxist concept of "Capitalocene" suggests; see Hamilton 2017: 29), or, for that matter, that geology has become philosophy.

And yet, the Earth System that is the new object the Anthropocene is all about, namely nature as impacted and altered, as unpredictably and erratically *changing* owing to our practices, and thereby in turn profoundly impacting and altering *us*, human society, in a dialectic interwinement of unprecedented vehemence, carries what Hamilton calls "ontological meaning" beyond its scientific importance. That it does so is a call for intellectual arms. It provides the point of departure for this book. It assigns philosophy with novel tasks and challenges its pre-Anthropocene presuppositions and concepts, removing the carpet under them with a fact-based, utterly concrete urgency for which most of what today passes as philosophy is, I am sorry to say, both ill-prepared and ill-equipped.

So what is this novelty about? What could possibly be its ontological meaning?

Again, only seemingly are these purely abstract questions, to be pondered the timeless way philosophers are wont to do, *sub specie aeternitatis*. The novelty that the Anthropocene consists in is multi-dimensional, of course. Yet it can be broken down to one instance of novelty in particular, one that very clearly questions philosophy as usual: without wanting to, we have through our activities, our technologies and economies, brought about a series of truly novel objects.

Radiation, hydrocarbons, and global warming are three such objects, leading Timothy Morton to coin the notion of "hyperobjects". What can be said about them? Massively distributed in time and space, they are nonlocal. They are objects "already there", before we direct our attention to them and acknowledge them, and so precisely not products of our intelligence and intentions but overwhelming them, dwarfing them, and so in no way a mere function of our measuring devices. Hyperobjects, the examples demonstrate, are a truly never-before-seen sort of object in that there is no "away" to which we can, say, sweep the radioactive dust or "do away with" the utterly concrete – physical, biological, chemical – effects of global warming. As genuinely nonhuman entities that, as Morton explains, "are not simply products of a human gaze", hyperobjects are huge objects consisting of other objects, as it were: "global warming comprises the sun, the biosphere, fossil fuels, cars, and so on" (Morton 2013: 48, 120, 199). In short, they are objects such that there is nothing *outside* of them, not comprised by them, impacted by them. They are total and totalizing (my words, not Morton's), reconfiguring whatever totalities (say, ecosystems) preexisted them, creating a situation where everything that comes to pass is either facilitated or rendered impossible (shut down, going extinct) given the weight of the ubiquity and inescapability of their presence: they will not go away, neither in the dimension of time nor in that of place.

Whereas Morton is spot on about the novelty of hyperobjects, he is so carried away by his own rhetoric that he ends up mystifying them, ignoring the sociopolitical forces that help produce them and the companies – fossil, capitalist, etc. – that are hell-bent on fueling the spread of hyperobjects instead of minimizing it so as to avert the dangers scientists warn against. Morton gets it wrong when stating in his recent book, *Being Ecological*, that "ecological facts are about the unintended consequences of anthropocentrism" (2018: 28). In the present situation and given the knowledge we have of the causes driving the process, speaking of "unintended consequences" is simply misleading, verging on intellectual dishonesty. Why?

To be sure, the ecological facts, in the shape of catastrophy that Hamilton rightly insists on, may have started out as unintended such, and as unforeseen. But to say, as of 2018, that the facts in question are unintended is to lend them a degree of innocence wholly unwarranted – epistemically, politically, and morally. No human actor has ever "intended" to bring about global warming. Granted. But there is a well-documented one-directional causal link between my buying the new Smartphone and the logging of what remains of pristine forests in Congo: the phone consists of "matter", a key component being coltan, to be found in this forest and nearly nowhere else. Buying the phone drives the deforestation that drives global warming – and about each and every step in this fairly simple causal sequence, we are *perfectly well informed.* Knowing what we are part of in doing what we do, we cannot disconnect our actions from their well-reported consequences; we cannot confine our responsibility to consequences that we wish for and deny it for, or feign ignorance about, the rest of them. For all the factual simplicity of the causal sequence are thus implicated in, however, we are pretty complex and full of contradictions, "hypersubjects" perhaps. Take the case of Morton himself, regarded in some circles as the world's leading ecophilosopher. By the end of 2016, as he wrote on his blog, he had racked up 350 000 air miles for the year, flying worldwide to so many conferences to share his eco-concerns. At the same time, he is reported to "become so emotional that he actually starts to cry" when discussing mass extinction (Blasdel 2017: 29).

Morton's notion of hyperobjects is instructive in another respect as well. It helps put into sharp relief how one of the most venerable of all philosophical distinctions since Plato and Aristotle is now being blurred, possibly to the point of dissolving completely: the distinction between the grown and the made, or, if you will, the organic and the artificial. Hyperobjects, such as those now inseparable from the reality of the Anthropocene, are clearly not *grown* in the Aristotelian sense of emerging organically in the natural world, as do the worm and the oak tree. Nor are hyperobjects (take radioactivity) *made* in the traditional sense of being brought into existence in a manner presupposing human agents who decide to go about making them, with a view to some perceived human interest or need.

Cosmologies in general, and philosophically elaborated ontologies in particular, have the task of explaining what is – meaning it all, *everything*, in such a way as to account for and do justice to all (property-based, qualitative) *differences* between what exists. As we will see, however, it is not only that the cosmologies that I will examine come up short when confronted with the novelty of hyperobjects – unsurprisingly so. The typologies typically offered by (Western) cosmologies also lack proper categories for the kinds of artifacts we literally litter our human existence on planet Earth with these days. Plastic is a good example. Humans have produced 8.3 billion tonnes of plastic since the 1950s, which is more in the last decade than in the entire twentieth century; a million plastic bottles are bought around the world every minute, with most ending up in landfill or the sea (Taylor 2018: 12). Microplastic, a miniscule product sufficiently potent to alter all the ecosystems of the seas, has now entered the entire marine food chain, scientists report. This is occurring as part of what is in itself a historically unprecedented situation in that, as of today, in Western societies, and increasingly in the whole world, virtually *everything* a child relates to – sees, hears, smells, tastes, and touches; in short, experiences first-hand, sensuously, bodily – is a human artifact of some sort or other, not something grown in and as part of the natural world. To the extent that there has historically been an asymmetry between the two, it has throughout history been of the opposite kind, with the grown vastly predominating over the made.

How serious a problem is it that new entities (to use the neutral, all-across-the-board term), at unprecedented pace and with unprecedented powers of disruption, emerge? For a start, where do we locate them – in culture or in nature? Are such entities a problem for our way of thinking about the world, exploding the concepts, categories, and distinctions that we are socialized into employing? Or, do they primarily pose a problem on the level of action?

If the Anthropocene is the historical product of anthropocentrism, it is also what forces us to abandon it and search for alternatives – alternatives whose first assignment is to be less destructive to the natural world that humanity depends upon: to help us learn, finally, to *appreciate that world for what it is in itself*, and to do so for other reasons than those linked to our obvious stake in securing the survival of humanity – admittedly something that we are not particularly good at, even though – or perhaps in part because – we are still committed to anthropocentrism. Indeed, in keeping with the urgency of the situation, to be worth its salt a philosophy for tomorrow's world needs to rethink the relationship between humans and the rest of nature in a way that helps us recognize the manifold of agents and agencies in beings other than humans. This undertaking is no less normative than the one it seeks to replace. Only its normativity is explicit not implicit, and – what matters most – it's of a substantially different kind in acknowledging value as well as agency in so many different beings and forms of life in nature. In rejecting the thesis of human exceptionalism on which anthropocentrism is premised, humans are a part of nature, not apart from nature. Not only have we been wrong about all those "others", taking them to be what we are not, and vice versa, but we have been wrong about ourselves as well.

There is an irony in this search for alternatives suited to let us grasp both what is at stake in the Anthropocene and how to avoid the worst in terms of the destructive impacts set in motion. The irony is betrayed in the need to "at long last" come to acknowledge the falsity of human exceptionalism and admit to non-human beings capacities we routinely deny them. Novel phenomena call for novel ways of thinking and coping, don't they?

Einstein was right that new problems cannot be solved within the framework that produced them. The irony I have in mind, however, is that instead of trying to come up with something truly new in order to match the novelty of the Anthropocene, we would do well to take a fresh look at what is widely considered as the oldest cosmology in human history, referred to as either animism or panvitalism, or – put philosophically – as panpsychism.

To dismiss as philosophically worthless everything that predated the Cartesian mind-matter dualism, taken as roughly corresponding to the division between culture and nature that we looked at above, would be to repeat just what did the thinkers that, more or less unanimously, have for the last four centuries promoted the very paradigm that the advent of the Anthropocene now tells us is spelling disaster. It turns out that we took the wrong turning, so let's consider what was cast aside as belonging to the dustbin of history so as to facilitate the path taken with such (it now dawns on us) disastrous complacency.

Panpsychism has been with us from the start, not only historically, but also in this introduction. The critical portrait of the mechanistic worldview with which I began, articulated by David Skrbina, is a portrait painted from a specific alternative point of view, that of panpsychism. Crudely put, the two views are to each other as thesis is to anti-thesis; no two doctrines could contradict each other more completely and exclude each other more fundamentally.

What exactly is panpsychism? I for one still wonder, despite having completed the chapters you are about to read. There is the easy part and there is the difficult one. The easy part is to do with the claim that everything that exists exhibits mind, by which is meant (in various degrees, from the primitive to the most advanced) mentality, interiority, intelligence, and purposiveness. To postulate as much on behalf of "everything" is, however, so radical a thought as to be beyond what we can ever hope to settle. Does a stone possess properties of the kind meant here? Is there anything mind-like about such a thing as a stone at all? Isn't there, in fact, only so many properties *excluding* everything we associate with mind, and so with agency, insofar as we take agency to presuppose and require mind – a capacity for thought – for intentionality, for the non-arbitrary pursuit of goals, and the like? The only answer for this side of rational argument, of what we expect from science, is probably the one Bertrand Russell came up with, though laconically so and clearly hoping for more: we will never really know.

Granted. But perhaps we will be able to know if we shift from entities as foreign to us as stones to entities with which we evidently have something in common as far as the relevant properties of mind-like character are concerned – for example, animals such as wolves or eagles (notwithstanding Descartes' view of animals in general as mindless "mere automata"). Yet, what happens in doing so is that the more plausible it now appears to us that we have indeed to do with a mind-possessing nonhuman creature (the wolf, the eagle, and so on), the more does the philosophical radicalism of panpsychism get lost – the entities *very different* from us on behalf of which panpsychism makes its bold claim is lost from view as we narrow the field under consideration to entities sufficiently like us to qualify (by us, at that) as possessing some degree of capacities that we pride ourselves in having "more of" than anybody else. Whether by fiat, bias, or default, the stone disappeared from the picture.

So even the easy part is not easy at all. The claim that something identical, or similar, holds for everything that exits is easy to grasp as a one-liner and is as general as they go. But we must admit it lands us in great difficulties the moment we try to assess its validity in a concrete way, applying the general idea to the particular cases in all their heterogeneity.

This is where the difficult part kicks in. The quotation from Skrbina is a good example: From his implicit panpsychist point of view, he starts by taking issue with the dominant orthodoxy that "inanimate things are fundamentally devoid of mental qualities". He goes on point out how we "treat nature as an impersonal thing or collection of things, without spontaneity, without intrinsic value". And he ends by stating that "natural resources, plant and animal species have been exploited" for human benefit.

Skrbina's point about our exploitative practices is well taken. But it is not what will be considered the truly contestable part of his position. The difficulty his statements illustrate is conceptual, and it is one that I will later expose in one affirmative version of panpsychism after the other. What is problematic about them is the tendency to commit the very slide that we encounter in Skrbina's paragraph: the slide from talking about "inanimate things" to talking about "plant and animal species", starting with the former, ending up with examples from the latter. The slide is symptomatic of an oscillation between entities (and I deliberately use the generic and general term) of very different kinds, so different that putting them on the same level, that of "everything" that allegedly exhibits mind or experience (not the same, we will see), risks ignoring or denying the differences that make a crucial difference between them. Only at the cost of gross simplification can we describe in like manner, using the same terms, so unlike entities as stones and gorillas, tables and oaks, molecules and brittlestars, creeks and peccaries, worms and dogs (my examples are not coincidental; I will return to each of them in later chapters).

That sweeping claims on a high level of abstraction get into trouble once we take them down to earth is nothing new under the sun, meaning that panpsychism is in good company. However, the problem is more acute in the case of panpsychism because its ability to say something more true and more precise than its rivals about the various entities existing on earth is what its chief validity claim consists in.

To be able to demonstrate as wrong the claim shared by the mechanistic worldview and the anthropocentrism wedded to it that "inanimate things are fundamentally devoid of mental qualities" is a tall order. To show the wrongness of the claim that plant and animal species are devoid of them is considerably less so. Again, in saying this it is tempting to declare the subject closed, in the sense that we have reached an absolute limit to what we can know, what we can hope to establish as true and valid knowledge about the subject matter. Concretely, we may establish knowledge about the mental qualities possessed (or not possessed) by a given animals species (i.e., can fish feel pain?). The more advanced the species (say, dolphins or chimpanzees), the more sophisticated are its mental capacities, and the easier it is for us to acknowledge them, since we take the animal in question to be "so much like us". Hence, they are able, in some measure or other, to engage communicatively with us (betraying that we are engaged in shamelessly anthropocentric comparison, and how could we not be?). But we may not, and likely never will, gain the same knowledge about the mental qualities of stones and molecules, or of creeks and peccaries. The reach of our well-proven knowledge about entities other-than-human remains limited, stubbornly and frustratingly so, and for that reason most of what we would like to illuminate in that vastest of realms, the natural world, must remain in the dark.

As will become clear in Chapter 1, dealing with Thomas Nagel's *Mind and Cosmos*, I am deeply dissatisfied with this limit: not with there being one, but with the tendency to premise the limit as we understand it upon the sole criterion of the reach of our knowledge. That is to put the cart before the horse: it is to perpetuate the privileging of what we humans can know over what exists and the myriad ways in which entities different from us exist. This *primacy of epistemology over ontology* is axiomatic and as such unquestioned in the mechanistic worldview and its ally, the anthropocentric way of assessing what the world consists of, namely in terms of means for human ends. Moreover, the reasoning that the primacy of knowing over being comes down to is logically untenable: it does not follow from humans' *not knowing* whether nonhuman beings of various kinds possess the same (largely intellectual) capacities that we do, that we may treat them as though they don't. Put differently, even if the philosophical consensus within the Western tradition is that we can never really know whether various animals and plants, or ecosystems such as forests and seas, may be said in a meaningful way to manifest agency (the way we humans define it and look for it), and whether they qualify as ends-in-themselves and as exhibiting intrinsic value, it does not follow from this knowledge deficit that the entities in question be treated as mere means, as devoid of what we find ourselves incapable of recognizing in them. If anything, that is *a deficiency on our part, not on theirs*, a deficiency located within our anthropocentric subject-centered realm, not within the natural world we pronounce on. The creatures that make up the latter should not be punished for a lack that reflects on us, not them.

Return now to Clive Hamilton's claim that the appearance of this new object, the Earth System as impacted by the advent of the Anthropocene, has "ontological meaning". What can that mean? Much of what I have to say in this book, especially in Chapters 4 and 5, is my response to this question.

It remains true that everything we may think, say, and do with regard to other-than-human entities – be it recognizing their intrinsic value and their intelligence, their agency in the form of actively adopting *their* distinct point of view *in*, and *on*, the world – we will think, say, and do from our human point of view within that larger world of which we are all part, meaning participants. But that is a hermeneutical point, born from a fact that we cannot escape. It marks our peculiarly human-based condition of possibility for grasping what exists and happens in that wider world, and in doing so draws up the limits to what we can know and understand.

We would commit a category mistake if from this fact we conclude that anthropocentrism is the only possible, and therefore the only valid, philosophy. The fact that there is a manifold of other-than-human points of view in/on the world, each relating in *highly distinct ways* to what comes to pass in it, each enjoying what is always a distinctly limited access to the world, namely the kind of access specific or even unique for *the sort of* entity or being in question, is not a reason to disqualify the validity and relevance of those other-than-human points of view. On the contrary, it signifies that there is *all the more reason to be attentive to them*, the more different they are from our point of view, selective and limited as that point of view is, inevitably so, just like everybody/everything else's. My critique of anthropocentrism is not about giving up on the human point of view. It is about situating it as one of many.

What does the novelty of the trajectory from anthropocentrism to the Anthropocene consist in? And how well equipped is a cosmology like panpsychism to account for this novelty, to do the sort of justice to it that other cosmologies cannot? Does the fact that panpsychism is the oldest *Weltanschauung* that we know of not count heavily against its prospects in grasping what is truly novel about our present era?

To anticipate, I do not end the cosmological journey that this book aims to be by concluding that panpsychism – in some particular version – provides all the answers that we are searching for in our confrontation with the Anthropocene. To be sure, there will be more questions asked than answers given; and for each answer proposed, new questions will arise. That said, one of the strengths of panpsychism, following from its preparedness to open up for and take into account so many other-than-human points of view in/on the world, lies precisely in the contrast it thereby marks with the worldview we for so long have deemed the only, or supremely, "rational" and "scientific one". In predating the worldview that paved the way for the Anthropocene, panpsychism may help us identify where we went wrong, having so clearly at some point taken the wrong turning. To echo what I argued in *Denial*: if a specific culture's view of nature accompanies and directly or indirectly justifies what in the natural world amounts to a series of practices downright destructive to that nature, then that culture and that theory (cosmology) cannot be right, cannot be true. They can't be true since they demonstrably fail when it comes to the bottom line of all practices and theories alike: securing survival, as opposed to jeopardizing it.

Remember this: before man was removed from nature, mind was removed from matter. The Anthropocene is about being, for everything that exists, at the receiving end of this twice over removal, one not merely thought but physically enacted. Again: a way of looking at the world and acting in it that leads to the decimation, degradation, and extinction now occurring, pointed out as so many matters of scientific fact by Ceballos et al. in the above quote, cannot be epistemically valid (correct about its subject matter), and cannot be morally right. The true relationship between culture and nature cannot be one of systemic destruction of the latter by the operations of the former. It must instead be one of enduring co-existence, not only compatible with a future life on planet Earth, but actively promoting and affirming it in its full richness, complexity, and heterogeneity, all across the conventional human-nature divide.

Since my approach highlights the *practices* that any given cosmology helps inform, legitimate, and render meaningful, with a particular view to how the natural world in its entirety is affected by those practices, I have found it instructive to engage with a discipline that all-too-rarely figures in philosophical discussion, namely anthropology. This means that the argument I first set out theoretically by way of engaging with Thomas Nagel and Alfred North Whitehead, becomes increasingly concrete – situated in specific cultures and natures – as the book proceeds. The movement enacted in the course of the chapters is one from the abstract to the increasingly more empirical: from ideas about the nature and extension of experience to what various sorts of beings in the contemporary world in fact experience, as part of their present situation, all too frequently in terms of danger – endangered survival prospects – at that. Once consulted, the relevance to my topic of ethnographic studies will, I hope, be as obvious as it is indispensable. Given my emphasis on practices, on how what a given culture in a given epoch induces its members to think about the world pervades their doings in it, the work done by anthropologists will lead the way in this book's centerpiece: Chapter 4 where I present my thesis that animism is panpsychism in practice.

What do I mean by suggesting that the relevance of panpsychism in the Anthropocene can only be fully appreciated if we open the philosophical discourse to the research carried out by anthropologists?

First, the best anthropological work on cosmologies – understood as a total view of reality, depicting the "larger-scale structure, origin and evolution of the concrete world" (Mathews 1991: 11) – is as philosophically profound as it gets. And whereas philosophers, analytic or not, typically take no interest in anthropology, the best anthropologists are true philosophers. Indeed, their subject matter demands nothing less. Exposing the view of reality that one has internalized as second nature to a cosmology of a genuinely different kind, studying how it is acted upon by a particular indigenous group, supplies a unique opportunity for critical reflection on behalf on one's own culture and the cosmology that it demands allegiance to.

Second, in discussing the work of leading contemporary anthropologists such as Tim Ingold, Philippe Descola, and Eduardo Kohn, the dead ends of the current debate in which the culture/nature distinction is being deconstructed (as the discourse has it) can be identified and (hopefully) avoided. In particular, a recent ethnographic study such as Kohn's *How Forests Think*, carried out among the Runa of Ecuador's Upper Amazon, helps us radically rethink the interconnection between culture and nature, and, as part of that – albeit not in the fashion of the one-to-one correspondence we as present-day Westerners are accustomed to – the inter-relationship between various sorts of agents, some of them human but many of them not, yet for all that commanding and performing the role of so many persons, in so many different forms, guises, and ways of being in the world and relating actively, from a perfectly distinct point of view, to everything and everybody it encompasses. We humans are indeed creators of a world (our world). But so are so many others as well, being subjects who sometimes relate to us as fellow subjects, who sometimes objectify us, demonstrating that such distinction-making is anything but an exclusively human capacity and concern, and so calling the bluff of anthropocentrism before our eyes.

Third, in keeping with my thesis that animism (studied by anthropologists) is panpsychism (theorized by philosophers) in practice, the ethnographic works I draw upon help vindicate what has for long been a forgotten, if not outright tabooed, insight: that we come into this world expecting it – all of it – to be alive. That the world is alive, that to be means to participate in this aliveness that is encountered everywhere and in everything, is the meeting-point between panpsychism as philosophy and animism as practice. The expectation that the world is alive and must be treated as such is also, no less significantly, what appears to be a universal feature among humans, innate in the newly born, acted upon as a matter of course by the child (an insight many a reader may have first encountered in Paul Shepard's classic *Nature and Madness*). I shall have plenty to say about how socialization into adulthood currently sees to it that this expectation be banished and given up as – exactly – "childish". Thus, we become alienated from its truth, so obvious and indubitable to humans and cultures everywhere in all epochs up to the present, showing us up as the utterly remarkable exception from the historical rule, having shifted from experiencing the world as alive to experiencing most of it as not-so, as non-agential, as "mere matter", in a turning of the tables that our predecessors everywhere would have deemed a proof of madness. To echo Shepard's book title, being, moreover, the very shift that inspired his book's unforgettable and to this day acutely pertinent opening question: "Why do men persist in destroying their habitat?" (Shepard 1982: 1).

In seeking to root out, tooth and nail, all signs of our innate affinity to nature and "childish" openness to communicating and interacting with everything encountered in the outside natural world, so-called anthropomorphism has been virtually banished. We thus have it backward: it is not that we are projecting a "subjective" liveliness and agency onto the so-called external world (taken as a blank slate); on the contrary, we are addressed by, and respond to, the liveliness and manifestations of agency that we encounter as the very fabric of the world. That this is so is not a timeless truth, however, immune to change. Indeed, the longer and more effectively we look upon and treat all those other-than-human agents as nothing of the sort, as objects not subjects, the more they will tend to become so. This follows the logic of self-fulfilling prophesies – deprived of the concrete conditions for their agency to thrive, becoming few and far between as a direct consequence of the Anthropocene's coinciding with the sixth mass extinction in history, unprecedented in being driven by human activities.

Fourth, insofar as the philosophical discussion of agency is itself changed by the event of the Anthropocene, an engagement with insights from anthropology allows us to see more clearly just where currently popular versions of posthumanism go wrong. Notwithstanding posthumanism's cogent point of departure – namely its misgivings about the whole anthropocentric framework, intellectually, culturally, and politically – what its leading proponents offer in its stead in terms of an alternative cosmology suffers from a series of blind spots, dead ends, and category mistakes. In Chapter 5, I single out Bruno Latour's recent *Facing Gaia* as a particularly instructive case in point. Put bluntly, what Latour offers is what I critique as a "flat ontology", one where the eagerness to claim agency and "agential" capacities on behalf of the myriad of other-than-human entities that are denied such in anthropocentrism. "Flat ontology" ends up painting a picture where all entities are grey, where one and the same portrait of agency-related capacities is brought to bear on virtually everything that Latour casts his eyes on, regardless of their differences. This means disregarding the very qualities of otherness (from us humans, but also from each other) that an ontology alternative to the dominant one needs to honor and to which it must do conceptual justice.

For partly related reasons, this criticism is one I also level at Karen Barad's thought-provoking *Meeting the Universe Halfway*, a book applying important insights from the physicist Niels Bohr to explore what it meant by the "entanglement" of matter, meaning, and experience. Common to Latour and Barad, and to a kindred spirit such as Jane Bennett (author of *Vibrant Matter*), is the collapsing and leveling of agent-specific differences that their attempts to vindicate them ends up committing, thereby undermining the promise to fully differentiate the various agencies in their present predicament of precarity, of demise and imminent extinction. The Anthropocene, I am sorry to say, is not about the latest fashion in philosophy, but about the all-too-mundane real effects of all-too-powerful, human-initiated and human-perpetuated practices in the modus operandi of systemic overshoot and exploitation; not about the proliferation of agencies, of letting hundred flowers blossom, but about their ongoing jeopardy and decimation. That is why it calls for mourning – a topic-cum-experience that the various posthumanisms are conspicuously silent about, a silence complicit in prolonging the dangerously abstract guise that the natural world has assumed in our culture.

1 Getting it right about mind, nature, and cosmos

This chapter explores the attempt by the philosopher Thomas Nagel to re-engage with the classical questions of philosophy in general, metaphysics in particular. He does so against a predominantly negative background: namely his profound dissatisfaction with the view about reality and about nature – broad labels for big topics, indeed – that he sees as enjoying hegemony within philosophy as well as the natural sciences. As the discussion proceeds, however, it will become evident that Nagel's attempt to work out a corrective to the present paradigm is far less radical and wide-reaching than he seems to be aware of. As later chapters will demonstrate, that this is so, and the reasons for it, is not peculiar to Nagel's contribution; it is a feature of several other attempts to break out of a declaredly prison-like human-centered deadlock as well.

I

Nagel tells his readers that the aim of his book *Mind and Cosmos* is "to argue that the mind-body problem is not just a local problem, having to do with the relation between mind, brain, and behavior in living organisms, but that it invades our understanding of the entire cosmos and its history" (2012: 3). In speaking of "our" understanding, I take Nagel to mean that of Western modernity, including laypersons as well as philosophers. He goes on to observe that intellectual humility requires that we resist the temptation to assume that "even the best developed and most successful forms of contemporary scientific knowledge" are in principle sufficient to understand the universe as a whole (ibid.). Whereas most practicing scientists according to Nagel do not express views about the natural order as a whole, among those who do, joined as they are by philosophers, "reductive materialism is widely assumed to be the only serious possibility" (4). Nagel's claim is that such materialism, which currently takes the form of psychophysical reductionism, fails in its aim to show "how the physical sciences could in principle provide a theory of everything" (ibid.). The common assumption that everything, including mind, is physical, if only we study it hard enough, must therefore be rejected. Indeed, it "flies in the face of common sense" (5). Alluding to the subtitle of his book, "Why the materialist neo-Darwinian conception of nature is almost certainly false", Nagel considers it "prima facie highly implausible that life as we know it is the result of a sequence of physical accidents together with the mechanism of natural selection" (6). The resistance of common sense to a fully mechanistic account of the origin and evolution of life, held to depend only on the laws of chemistry and physics, is perfectly legitimate. There is no reason, Nagel asserts, to hold the reductive research program as sacrosanct, although that is precisely what "almost everyone in our secular culture has been browbeaten" into doing (7).

Coming from one of the world's most renowned analytic philosophers, Nagel's rhetoric is unusual, and surprisingly strong-worded. In presenting himself as a "layman who reads widely in the literature that explains contemporary science to the nonspecialist" (5), and in siding so explicitly with common sense, Nagel's project in *Mind and Cosmos* should be understood as a wide-ranging attack not only on philosophical positions – typically, reductionisms – about which common sense is probably perfectly, and happily, unaware. The book is also meant to challenge the way hegemonic science as well as philosophy have pervaded culture, alienating common sense in its quest for convincing answers to the fundamental questions about the origin of life as well as life's meaning and value.

Here lies a chief claim of Nagel's: As humans, we aspire to being able to understand reality in its totality. Nothing less will do. This aspiration is a given for Nagel, indubitable and ineluctable, helping define the human condition. We crave to understand the universe, the natural order, everything that goes into making, and into making possible in the first place, the world of which we are part. It follows that the explanation has to be intelligible: as intelligent beings, creatures endowed with mind, we will not settle for less. Hence, for an explanation of the world to be complete, and to qualify as valid, it has to include at its core an account of *ourselves*, doing justice to the intelligence that we enact in raising the question about the origin and nature of the world, of everything we take to be real. In enquiring about the world, we also enquire about ourselves; we are epistemic subjects asking questions and demanding coherent answers about epistemic objects.

The reductive materialism that explains the appearance of life from dead matter and that takes evolution to involve nothing but the operation of physical law is not an unassailable doctrine, let alone the one and only scientific theory, says Nagel. Rather it reigns by default. There is something deeply problematic about a theory that seeks to give a complete explanation of the universe and its entities through their unification, by reference to what each and every one of them is taken to have in common. If different entities – some living, others dead, some animate, others inanimate – are said to consist of the same element, how do you account for their difference, for what distinguishes one from the other and marks its specificity? Reductionism, in whatever guise it is conceived, may have the virtue of identifying something common in the manifold of things existing. But it has the vice of losing out on what bestows an entity's identity as *this* entity as opposed to that one. In a word, reductionism fails to account for the ontological richness and diversity of the world.

Admittedly, and as anticipated in my Introduction, these are my concerns more than Nagel's, to be more fully developed later. Nagel's way of raising the issue of diversity – understood as *irreducible such* – takes the wholly conventional form of asking how mind (or the mental) can possibly emerge from something purely physical. If you posit the physical as what indubitably exists, and with everything existing as physical (in some minimal sense), how do you explain the existence of mind? More specifically, how do you do so given that – as common sense would insist – mind is something that the "merely" physical is not, that mind can do something, can accomplish things (ask questions, entertain ideas, ponder contrafactuals, etc.) that physical things demonstrably cannot?

"A genuine alternative to the reductionist program", writes Nagel, "would require an account of how mind and everything that goes with it is inherent in the universe" (15). Perhaps, he goes on to ponder, "the natural order is not exclusively physical" (16).

To suggest that mind is inherent in the universe is certainly to challenge all-out the materialist doctrine. We should, however, note something that Nagel tends not to, namely that his suggestion is not necessarily incompatible with reductionism; on the face of it, it may be a case of replacing one *type* of reduction with another, that is, a physical one with a mind-oriented one. But then again, to say that mind is inherent in the universe need not imply that *everything* existing there has mind-character. Rather, it may mean that the existence of mind is to be seen as an ineliminable component of the universe, yet not to be found in *every* entity but instead in some or in most of them.

However intriguing, Nagel's suggestion, here as elsewhere, inherits the trait of excessive generality from the doctrine that it seeks to reject. There is simply, on both sides of the disagreement, a lack of a viable way to settle the matter: to show, beyond dispute, whether the universe, or the so-called natural order, in its entirety contains physical, meaning nonmental, properties, or whether it – all of it, as Nagel seems to suggest in his cited formulation – contains mental ones, called mind. As we shall see more fully below, these options by no means exhaust the possibilities; but they are the ones most suited to illuminate Nagel's suggestion.

A crucial step in Nagel's argument in *Mind and Cosmos* is his statement that his "guiding conviction is that mind is not just an afterthought or an accident or an add-on, but a basic aspect of nature" (16). This of course recalls one of the standard arguments against physical reductionism: that it cannot account for mind, and that its attempts to do so amount to a series of *ad hoc* maneuvers so as to add something (the specific qualities of mind) that is fundamentally alien to it. More important for my purposes is the way Nagel seeks to back up his "conviction", namely by contending that regarding mind as a basic aspect of nature "is one of the background conditions of science". And science is "driven by the assumption that the world is intelligible". Nagel's idea here is worth quoting in full:

> The world in which we find ourselves, and about which experience gives us some information, can be not only described but understood. That assumption is behind every pursuit of knowledge, including pursuits that end in illusion. In the natural sciences as they have developed since the seventeenth century, the assumption of intelligibility has led to extraordinary discoveries, confirmed by prediction and experiment, of a hidden natural order that cannot be observed by human perception alone. Without the assumption of an intelligible underlying order, which long antedates the scientific revolution, those discoveries could not have been made.
>
> (16)

We have already encountered Nagel's notion that, being intelligent, the understanding of the world – including nonhuman entities – must be such as to seem intelligible to us. The constellation with which Nagel works is that of us as knowers and the world as knowable to us. The world is knowable when it matches up to our faculties of knowledge: reaching out to what we expect to qualify as intelligible, we settle for nothing else.

This notion of Nagel's may seem sensible enough, even obvious. My worry, however, is that Nagel reduces ontology to epistemology: he reduces the metaphysical question of *what is* to questions about how we can come to *know* what is. As we will see in a moment, there is a slide in Nagel's book from start to end. He begins by asking the big questions about the big picture, meaning our modern conception of nature, understood as the totality of what is. Hence, he aspires to reopen the classical metaphysical inquiry into everything that, in one form or another, makes up reality. The aim of this inquiry is, and remains, totality, an exhaustive theory of what is, doing justice to the richness of everything that exists, yet does so in so many peculiar ways: the many beings that make up Being, to allude to Heidegger.

The slide that I see in the course of Nagel's discussion is a shrinking, a narrowing of perspective. It is a shift of emphasis from the big picture that is concerned with everything there is, "the natural order" of which humans are just one of many entities, to an increasingly one-sided concentration on the question about how to give the best philosophical or scientific account of what we humans are, especially in terms of consciousness and cognition – i.e., how can we explain that we have come to possess mental capacities of the sort required for asking questions about anything at all? Capacities of such power and sophistication as to seem unique to us among all beings?

It is tempting to respond by saying that the slide is perfectly logical: who would give that exhaustive theory if not the epistemic subject? Isn't it elementary that in offering a view – any view – about the totality of what is, we need to take into account *who* it is that offers that view, so as to be able to decide about its truth (validity)?

The problem is that Nagel, in proceeding in said manner, ends up offering his readers a critique of the modern Western notion of the so-called natural order that is both less deep and less far-reaching than it should have been, given the book's stated aim of re-engaging with totality. It is precisely the obviousness, the "how could it be otherwise?", of Nagel's discussion that betrays it, that gives away how conventional it is, how loyal it is to the paradigm he sets out to question, the one inaugurated by Descartes and brought to unprecedented level of "transcendental" sophistication in the epistemology of Kant.

The point I am making against Nagel is reminiscent of the general point Charles Taylor has made on numerous occasions, concerning what Taylor calls *the primacy of the epistemological*, understood as "the tendency to think out the question of what something *is* in terms of the question of how it is *known*" (Taylor 1995: 34). Taylor sees this as "a pervasive feature of modern intellectual culture" (ibid.). Notwithstanding attempts to get rid of it, the Cartesian legacy proves persistent, and I think Nagel is a case in point. Its gist is the insistence that "we can somehow come to grips with the problem of knowledge, and then later proceed to determine what we can legitimately say about other things: about God, or the world, or human life" (Taylor 1995: vii). In other words, we first need to be clear about the nature of knowledge, hence about ourselves, about our capacities as epistemic subjects, and about what it is to make a defensible claim about the world, or some particular entities in it.

Coming at the issue from a different angle than Taylor, namely that peculiar to his version of realism, Roy Bhaskar designates the tendency to "analyse being in terms of knowledge" the "epistemic fallacy", involving as he sees it "the denial of the possibility of a philosophical ontology". It is, Bhaskar argues, "a necessary condition for the occurrence of science that the world exists and is of a certain type. Thus the possibility of our knowing is not an essential property, and so cannot be a defining characteristic, of the world". Accordingly, "knowledge follows existence, in logic and time; and any philosophical position which explicitly or implicitly denies this has got things upside down" (Bhaskar [1975] 2008: 38, 39).

I am not suggesting that Nagel is a case in point for getting the relationship between being and knowledge upside down the way that Bhaskar describes. But Nagel's project would have gained in clarity had he acknowledged the force of Bhaskar's (as well as Taylor's) insight. Short of that, Nagel describes the assumption about the intelligibility of the world, that "rational intelligibility is at the root of the natural order", as a form of "the principle of sufficient reason – that everything about the world can at some level be understood, and that if many things, even the most universal, initially seem arbitrary, that is because there are further things we do not know, which explains why they are not arbitrary after all" (17). Nagel is in no doubt that the intelligibility of the world – first described as an "assumption" – "is no accident". On this view, mind is "doubly related to the natural order. Nature is such as to give rise to conscious beings with minds; and it is such as to be comprehensible to such beings. Ultimately, therefore, such beings should be comprehensible to themselves" (17).

This reasoning illustrates the slide that I referred to. The problem is not Nagel's interest in mind as a something that exists in the natural order. What is problematic is what he immediately proceeds to do with mind: first, connecting it to "conscious beings", second, postulating that nature is (must be) such as to be comprehensible to those beings, and third, concluding ("therefore") the comprehensibility to themselves of such beings.

Nagel here actualizes the point made, in somewhat different manner, by Taylor and Bhaskar. In what is meant to strike the reader as a perfectly logical reasoning, and as compelling, Nagel moves from talking about the natural order – the totality, comprising everything, capturing the agenda of classical metaphysics – to concentrating on the issue of comprehensibility. He focuses on comprehensibility not in a universal sense but as it arises for *one* being in the world of beings, *one* entity in the myriad of such [entities] within the natural order, namely us humans in our capacity as the ones seeking knowledge and asking about its scope and truth, as brought out in Nagel's formulation about the things we know and those we (up until now, at least) don't know.

What I am faulting Nagel for is that he drops the broad concern with "the things", with totality, in favor of the all-too-familiar, yet restricted concern with knowledge ("comprehensibility") and the capacities, especially mental capacities, of the sort of beings for whom such knowledge is central.

There is also another restriction involved. Why approach nature, as does Nagel, with a special view to how it gives rise to *conscious beings with minds*? Why should this be what is most important, and most interesting, about nature? In giving pride of place to mind, does one not risk ignoring other properties, and other forms of existence, in that nature? Why accord such privilege to the former? And last but not least, the million-dollar question: Does mind exist in humans only? If so, how widely and differentiatedly is mind present, or spread out, in the natural order as Nagel understands it?

As I have indicated, there is something question-begging and tendentious about Nagel's way of going about raising these issues. He tends to pose them in a restricted manner that narrows the scope of his inquiry as compared with the totalizing and exhaustive ambition of metaphysics. This is an ambition that has generally been given up since Descartes and Kant and the "turn inward" (Taylor 1989) that they, in their distinct ways, inaugurated, with the still-lingering consequence that the concerns of epistemology have taken center stage and eclipsed a full philosophical take on the whole of which humans are only a part.

II

Return now to Nagel's main argument. He dismisses the view that everything that exists can in principle be explained by the laws that govern the physical universe. Physicalism runs up against an insoluble problem: it cannot convincingly account for the existence of mind in the physical universe whose basic elements that it seeks to explain. Indubitable as the existence of mind is (at least to us, and that is what counts on this model), it remains an anomaly in physicalism: something it is forced to acknowledge and so to explain, yet it cannot. Having dismissed materialism, understood as the attempt to explain mind as a consequence of physical law, Nagel goes on to dismiss its traditional chief rival as well, theism taken as the attempt to makes physical law a consequence of mind. None of them succeeds as a comprehensive world view, Nagel maintains, adding that his "interest is in the territory between them" (22). More precisely, his interest is in a secular conception that "acknowledges mind and all that it implies, not as the expression of divine intention but as a fundamental principle of nature along with physical law" (22).

Having at the start of his book rejected the view that mind can be explained by, and can be accounted for as emerging from, something purely physical, it is not surprising that he opts for a theory of reality that allows for both something physical and something mental. Both must be accounted for within the same comprehensive theory. *Neutral monism* is name of the position he advocates, favoring it over the traditional alternatives of materialism, idealism, and dualism (5). (It is noteworthy that Nagel only mentions idealism in passing, offering no discussion of Hegel and passing over Kant in total silence.)

So what does Nagel mean by neutral monism? Revealingly, Nagel defines it as "an answer to the mind-body problem" in that monism holds that "certain physical states of the central nervous system are also necessarily states of consciousness – their physical description being only a partial description of them, from the outside, so to speak" (57). I find this definition revealing because it highlights what I discussed above, namely Nagel's tendency to move directly from the totality that a "comprehensive theory" aspires to account for, to engaging with the mind-body problem and states of consciousness. True enough, this should not surprise us since the title of this chapter in Nagel's book is "Consciousness". The question, however, is whether the move to concentrate attention on the mind-body problem and on how to account for states of consciousness is justified, given that what Nagel tries to advocate is his favored position regarding the metaphysical question about reality taken in its totality. Of the "things" (Nagel's term) comprising that reality, many do not have bodies, and many do not have bodies in some combination with mind. It hardly needs saying that the one kind of thing that does possess both body and mind is the human being; but that is only one among so many other things existing, in ways that call out for the interest and ability of a favored position – among alternatives within metaphysics – to do them explanatory justice. So, whereas the neutral monism that Nagel endorses aspires to a comprehensive view and explanation regarding everything, in actually arguing its case, Nagel concentrates on what holds for *one* being within the vast natural order that contains so many different ones, other ones. Is it too vulgar to see this as a case of excessive interest in ourselves, at the expense of others? Is it too vulgar to see this as a narrowing of scope that comes with heavy real-life and far from exclusively intellectual costs, as the advent of the Anthropocene now makes clear, to recall my Introduction?

The monism he advances will be universal, Nagel assures us. "Everything, living or not, is constituted from elements having a nature that is both physical and nonphysical – that is, capable of combining into mental wholes" (57). This is important in that it expands the scope that was narrowed in Nagel's earlier formulation. For all that, he calls it a "reductive account" and says that it can also be described as a form of panpsychism, taken as the position that "all the elements of the physical world are also mental" (57). He has a footnote that states that Alfred North Whitehead's philosophy of organism is a case in point. But he abstains from any further reference to Whitehead.

I shall come back to Whitehead's philosophy and its importance as a leading twentieth-century statement of panpsychism, the doctrine that everything that has physical existence also has a mental dimension, an inside as well as an outside, and an ability to relate to itself as part of such interiority. What matters now is Nagel's assessment of panpsychism, representing what should appear as a promising position given Nagel's rejection of physicalism, materialism, and theism.

Nagel has discussed panpsychism on several occasions, but in light of his wide-ranging agenda in *Mind and Cosmos*, the treatment he gives it is surprisingly short. The question he takes panpsychism to try to answer has to do with the development of organisms with a subjective point of view (equivalent to what I referred to as interiority above). Such organisms, Nagel suggests, must be held to have been there from the beginning, presumably all the way down, as it were, in keeping with Nagel's take on panpsychism as yet another "comprehensively reductive conception" (61). Nagel proposes the following comparison: whereas particle physics renders the character and cosmological history of the nonliving material world intelligible, the panpsychist claim that protopsychic properties are to be found in all matter is one that is in principle impossible to prove: we will never really know whether it is true or not. Protopsychic properties are "postulated solely because they are needed to explain the appearance of consciousness at high levels of organic complexity" (62). The problem is that they are indescribable, in contrast to the physical properties of electrons and protons. And since we have no foolproof way of making sure that protopsychic properties do exist in all matter, so that the highest levels of consciousness (presumably to be found in humans) can be explained as emerging from the lowest, and the evolutionary most recent from the oldest and most primitive, it follows for Nagel that panpsychism fails to convince.

Nagel's way of putting his criticism says a lot about how he conceives of panpsychism:

> Without something unimaginably more systematic in the way of a reduction, panpsychism does not provide a new, more basic resting place in the search for intelligibility – a set of basic principles from which more complex results can be seen to follow. It offers only the form of an explanation without any content, and therefore doesn't seem to be much of an advance on the emergent alterative.
>
> (62)

That said, Nagel grants that the proposal for panpsychism is not empty, taken as a sort of monism that attempts to "recognize the mental as a physically irreducible part of reality while still clinging to the basic form of understanding that has proved so successful in physical theory" (62). On the negative side, "the idea of reducing the mind to elementary mental events or particles seems unnatural in a way that physical atomism doesn't" (62). We have no idea of a "part-whole relation for mental reality", Nagel continues, that would qualify as equally clear and compelling as what pertains to the space-time framework of the physical world; no idea, that is, "how mental states at the level of organisms should be composed out of the properties of microelements, whether those properties are similar in type to our experiential states or different" (62).

These quotes aptly show what strikes me as the leading premise for Nagel's discussion: that panpsychism is modeled on physical theory, so that the sort of reduction to which physical atomism is committed can be used to decide its (panpsychism's) success or failure. The premise being tacit in Nagel, the question that he does not answer is whether the assumption that panpsychism can, and indeed should, be modeled on physical theory, is warranted. If Nagel's proclaimed dismissal of physicalism is seriously meant, why grant physical theory the status of model in this way? And more specifically, why invoke reduction as (presumably) at work in physical *atomism* to assess the panpsychist proposal for an explanation of everything existing? Doing so sits uneasily with the fact that proponents of panpsychism typically endorse *holism*.

To be sure, some of Nagel's formulations go in the latter direction, for example his observation that "if a universal monism is correct, it would mean that [the] psychophysical connections are unbreakable: one cannot have the mental without the physical aspect, or vice versa" (63). Overall, however, Nagel's discussion of panpsychism is marred by his tendency to invoke positions that he elsewhere is busy critiquing – materialism and physicalism – as the scientific gold standard against which the panpsychist alternative is being measured. This amounts to an asymmetry of burden of proof as between the rival positions in question that I find puzzling – including the fact that Nagel does not reflect on it. In this respect as in others, therefore, he appears much less of a critical questioner of the current paradigm than he purports to be, as evidenced in his remark that "a hypothetical monism that has expanded to encompass the mind is far more speculative [than the standard materialist explanation of both the living operation and the evolutionary history of physical systems]" (64). Why precisely is panpsychism in this respect far more speculative than standard materialism? It is so, Nagel asserts, "since it says only that there is more to the basic substance of the world than can be captured by physics and chemistry" (64). True enough, but wasn't one of Nagel's stated aims to dismiss both physics and chemistry as exhaustive accounts of reality and to look for an alternative that goes beyond them, an alternative that, viewed from the standpoint of physics and chemistry can hardly avoid being dismissed on grounds of seeming unduly "speculative"?

Obviously, to show that consciousness is a physically irreducible feature of certain organisms, or even more basically, that mind or the mental is as much in all entities that exist, is an extremely tall order. It therefore seems justified that Nagel closes his discussion of panpsychism by raising the following question, meant as one that, at the present stage at least, cannot be convincingly answered: "Why should those properties [of consciousness] make the appearance of such organisms, starting from inorganic matter, at all likely?" (65)

For all its suggestiveness, there is in my view something question-begging about this way of raising the issue. The difficulty of coming up with a convincing answer hinges on the notion of "inorganic matter" that Nagel employs, though without explaining it. For lack of precise definition, we are left to conclude that any doctrine that hopes to demonstrate that the property consciousness (or mind) is, in some manner and combination, co-present in inorganic matter, is doomed to failure, for the simple reason that *inorganic matter is just too ontologically different from everything we mean by the mental* (be it consciousness or mind) to be suited to co-exist with it, *especially* in the strong sense intended by panpsychism. Note that whereas Nagel faults reductionism for not succeeding in accounting for human mind (mind as present in humans), a panpsychist's complaint would be instead that reductionism fails to account for mind as present in *other-than-human* beings.

At this point many readers would have expected Nagel to confront Galen Strawson's defense of panpsychism, surely the most widely discussed one during the last decade. Strawson's argument can be paraphrased like this (following Simons 2006: 146; Strawson 2006):

1 We cannot deny the existence of experience.
2 Experience appears to be emerge from physical phenomena that are not themselves experiential.
3 Wholly non-experiential phenomena are not by their physical nature capable of giving ride to experience.
4 Therefore, either experience emerges magically from wholly non-experiential phenomena or the physical phenomena from which experience emerge are in some way themselves experiential.
5 Magical or brute mergence is absurd.
6 Therefore, the physical phenomena from which experience emerge are in some way themselves experiential (Micropsychism).
7 It is implausible to suppose that nature is so fragmentarily constituted that some physical phenomena are experiential while others are not.
8 Therefore, all physical phenomena are in some way experiential.
9 But all phenomena are physical (Physicalism).
Therefore, all phenomena are in some way experiential (Panpsychism).

Why does Nagel not engage with Strawson's argument for panpsychism, an argument that has attracted enormous attention, including an anthology containing seventeen commentaries on it by prominent philosophers?

It is a missed opportunity. It would have given Nagel a chance to point out what I find remarkable about the way panpsychism is advocated by many of his colleagues, namely as a case of what I would characterize as *sheer logical coercion*: you start by postulating a premise that everyone must agree to, posited as what we "cannot deny" (i.e., the existence of experience); then you proceed to infer a series of "therefores", each one said to lead by logical necessity to the next, ending in a conclusion to be regarded as just as compelling as was the initial premise (i.e., therefore all phenomena are in some way experiential).

Is there not something contradictory about showing that "all phenomena are in some way experiential" by way of recourse to *a purely abstract logical reasoning*? Does the issue – the presence or absence of experience in everything making up physical reality – not demand a role for actual *experience*, as opposed to a sheer *intellectual* capacity to acknowledge the presence claimed; this being, moreover, a capacity exhibited purely on the side of the (ahistorical, context-free, cerebral, body-less) subject performing the reasoning, and so without the subject being a concrete *someone* engaged in some particular relationship with *those non-human physical entities* whose nature the reasoning is all about? How can you establish the indubitability of experience by doing without it, bracketing it so as to leave its presence/absence to logic alone, carried out in an experiential vacuum, a no-man's land?

III

Let us take a look at Nagel's other works on panpsychism to see if they allow for greater clarity about his assessment of it.

In the much-quoted paper "Panpsychism", published in *Mortal Questions*, Nagel sums up his discussion by saying that "panpsychism should be added to the current list of mutually incompatible and hopelessly unacceptable solutions to the mind-body problem" (1979: 193). This recalls my earlier comment that exposing panpsychism to the so-called mind-body problem may not serve either of them well. That the mind-body problem has for centuries been widely seen as insoluble, or as one of the most difficult philosophical problems of our time, says more about the Cartesian dualism that gave rise to it and whose stubborn legacy continues to give it pride of place than it says about panpsychism. It appears an odd and restricted way to put such a metaphysical doctrine as panpsychism to the test – an oddity seldom recognized as such to the extent that Cartesian dualist premises continue to be pervasive, both culturally and within analytic philosophy.

In his paper, Nagel's more specific criticism of panpsychism comes out in his statement that "it is difficult to imagine how a chain of explanatory inference could ever get from the mental states of whole animals back to the proto-mental properties of dead matter. It is a kind of breakdown we cannot envision, perhaps it is unintelligible" (1979: 194). But what precisely is "dead matter"? When Nagel points out how little we "know about how consciousness arises from matter in our own case and that of the animals in which we [contra Descartes] can identify it" (1979: 195), his frustration is of course legitimate – we wish we knew more than we do. However, this hardly amounts to a strong counter-argument about panpsychism.

In his next treatment of panpsychism, in the book *The View from Nowhere*, Nagel seems to take a more open and positive approach. "The strange truth", he tells us, "seems to be that certain complex, biologically generated physical systems, of which each of us is an example, have rich nonphysical properties" (1986: 51). This truth will have to be accounted for by an integrated theory of reality, one that, however, will probably not arrive "for centuries"; yet if and when it does so, it will "alter our conception of the universe as radically as anything has to date" (ibid.). He then adds that he is "pessimistic" about the capacity of physics to provide a complete understanding of reality. Nagel's dismissal of physics in this respect is more strongly worded here than in his more recent *Mind and Cosmos*: "To insist on trying to explain the mind in terms of concepts and theories that have been devised exclusively to explain nonmental phenomena is, in view of the radically distinguishing characteristics of the mental, both intellectually backward and scientifically suicidal" (1986: 52). The final nail in the coffin is the assertion that "the difference between mental and physical is far greater than the difference between electrical and mechanical" (ibid.). Something like "the generation of mind out of the recombination of matter" appears impossible to grasp, and we will need "entirely new intellectual tools" to be able one day, if ever, to do so (ibid.).

I fully endorse Nagel's commitment to the Aristotelian principle that different subjects (or objects) require different methods, concepts, and tools to be adequately grasped. Method must never take primacy over subject matter; the latter should always dictate the former. So there is no reason to quarrel with Nagel's insistence that the difference between mental and physical commands that the intellectual tools devised for the one not be applied to the other.

For all its laudability in principle, however, the question is whether this Aristotelian precept is really suited to the case at hand. What we need to hear more about in a substantial as opposed to methodical and epistemological respect is *how* different are the physical and the mental really? What grounds, ontologically speaking, their difference(s)?

On one level, we know what Nagel will answer: We do not know, and most probably we never will. But why exactly is that? Is it because both the mental and the physical, in their respective uniqueness, are so hard for us know? Or is it because one is more difficult to know than the other? If so, which?

To shed light on Nagel's position on this question, we shall consult his essay "The Psychophysical Nexus", which first appeared in 2000 and to which Nagel refers his readers in his all-too brief treatment of panpsychism in *Mind and Cosmos*.

Unsurprisingly, the essay is presented as an attempt to explore the mind-body problem, namely by pursuing an approach that "takes the subjective phenomenological features of conscious experience to be perfectly real and not reducible to anything else" (2002: 194). Again, the uniqueness of the subject matter at hand (conscious experience) is unequivocally, and laudably, acknowledged. But what, one wonders, about the rather vague "what" to which such experience is said to be not reducible?

In keeping with current debates, Nagel shifts from the mind-body to the mind-brain case. He observes that "what happens in consciousness is pretty clearly supervenient on what happens physically in the brain", only to add that the present state of our conceptions of consciousness and neurophysiology is such as to render "completely mysterious" the strict dependence that must be regarded as a "brute fact": "If the physical necessitates the mental, there must be some answer to the question *how* it does so. An obviously systematic connection that remains unintelligible to us calls out for a theory" (2002: 207, 208).

Deliberately or not, this way of putting the problem recalls Descartes' endeavor to explain the interaction between mind and body in the case of humans. Whereas interaction is a fact that we cannot deny, the Cartesian dualist metaphysics according to which there are two independent and self-sufficient substances cannot convincingly explain their interaction. Nagel of course rejects Cartesian dualism, citing the causal role of mental events in the physical world (2002: 203). Why did

Descartes get it wrong? He took it to be self-evident, just by thinking about it, that the human mind was not an extended material thing and that no extended material thing could be a thinking subject. But this, Nagel remarks, is not nearly so self-evident as that a number cannot have parents. What does seem true is that "the concept of a mind or of a mental event or process fails to plainly leave space for the possibility that what it designates should turn out also to be a physical thing or event or process", the trouble being that "mental concepts don't obviously pick out things or processes that take up room in the spatiotemporal world to begin with" (ibid.).

At this point in an argument that, so far at least, is predominantly negative, one would expect Nagel to consider a full-fledged alternative to the said impasse. Having rejected physicalism, taken to eliminate what is distinctive and undeniable about the mental, as well as Cartesian dualism, which options remain?

Nagel tells us that he has occasionally been drawn to "some kind of property dualism", finding however that, like Cartesian substance dualism, "it seems just to give a name to a mystery and not to explain anything" (2002: 203). Here is why: "Simply to say that mental events are physical events with additional, nonphysical properties is to force disparate concepts together without thereby making the link even potentially intelligible. It suggests pure emergence, which explains nothing" (ibid.).

I propose that the following three features are particularly important with regard to what kind of theory Nagel is after.

First, Nagel concentrates on the *human* case, even though the philosophical subject he raises at the outset – how to explain the natural order – includes a vast number of other cases, including cases pertaining to how something physical and something mental co-exist and interact. Whereas for Descartes the paradigmatic case of the two-substance doctrine was the human agent, combining the two-in-one, mind and body, as properties of one creature, for Nagel the paradigmatic case of the "psychophysical nexus" is the mind-brain case as instantiated in humans.

Second, Nagel shares with analytic philosophy an obsession with the reliability and adequacy of the concepts being employed in order to account for the phenomena that we take the world to consist of: understanding the world and its ontological structure is necessarily something mediated, never direct, never raw; as mediated it is a conceptual undertaking, in the Kantian sense that the concepts we employ are not merely required tools but help constitute the world: no concepts, no object. Hence Nagel's move, in the last quote, from events to concepts: from the phenomena themselves, to the concepts we use to get at them, there being no other way to get at them, not for us. This raises the question whether Nagel would accept or reject Bhaskar's point from realism that "knowledge follows existence, in logic and in time", on whose acceptance Bhaskar, for one, hinges the possibility of a philosophical ontology.

Third, in asserting that "pure emergence" explains nothing, Nagel does not spell out which of the two logically possible cases he has in mind: whether it is the case where something mental is said to emerge from something nonmental (i.e., purely physical), or the other way around. I take it that it is the former, in keeping with the way emergence is currently discussed. But what justifies assuming *that* asymmetry, that order of existence-cum-appearance? What justifies postulating the physical as what exists most elementarily, and indubitably so, casting the existence (and distinctness) of the mental as somehow secondary, or derivate, and mysterious at that? Did Descartes not posit *mind* as what most indisputably exists? But if Descartes was right about that, what about the way he restricted mind: exclusively to the first-person human subject, raising doubt about the existence of mind not only in nonhumans including all sorts of animals but in *other* humans as well? Indeed, what is the most intractable problem left by Descartes – the mind/body problem or the problem of other minds?

I consider that each of the three highlighted features of Nagel's approach would have been seriously challenged had Nagel engaged with the position that Alfred North Whitehead elaborates in his *Process and Reality*. The "property dualism" Nagel mentions, only to quickly dismiss it the way we saw, is perhaps the closest he comes to alluding to Whitehead in "The Psychophysical Nexus". In Nagel's *Mind and Cosmos*, the one mention made of Whitehead is in a footnote noting (not very informatively) that his philosophy of organism "also belongs in the company [of panpsychism]" (2012: 58).

It is telling that in Nagel's paper "Panpsychism", his most systematic treatment of panpsychism, the one philosopher (besides Saul Kripke) whose work he dwells upon is Ludwig Wittgenstein and his *Philosophical Investigations*. The questions Nagel asks there pervade his entire output and will be familiar to everyone who has heard about his most famous text, "What is it like to be a bat?", namely: "If the occurrence of a subjective experience is not the possession of a property by *something*, what is it? And what connexion does it have with the organism" (1979: 190). Turning to Wittgenstein to settle these questions, however, hardly brings any progress, Nagel finds, insofar as on Wittgenstein's view "there is almost nothing to be said about what property must be possessed by what thing" (ibid.). Still, Wittgenstein is positive that mental phenomena must count as real, suggesting the following symmetry: just as mental processes have physical criteria, physical processes have mental (specifically observational) criteria. Only what permits public agreement qualifies as real, according to Wittgenstein, and mental processes satisfy this requirement just as do physical ones.

What leaves Nagel less than satisfied, though, is that Wittgenstein's view depends too heavily on language, human language at that, I should add. Wittgenstein's account of mental phenomena deals with how they are ascribed, particularly in the first person. The limit of scope is obvious. As Nagel says: "But not all conscious beings are capable of language, and that leaves the problem of how [Wittgenstein's] view accommodates the subjectivity of *their* mental states" (1979: 191). As things stand, we are stuck with concepts about mental processes and about experience that are anchored in their application to humans. Nagel's dissatisfaction is to do with his conviction that "the experiences of other creatures are certainly independent of the reach of an analogy with the human case. They have their own reality and their own subjectivity" (ibid.). That sounds fine. But how can Nagel *know* that this is so? Does he actually claim to know as much, and to be able to impress upon others the validity of such knowledge, if we require that it be a knowledge (truth) to be had by all, to be public in Wittgenstein's sense, and not exclusive or private?

Note that I am not faulting Nagel for asserting that other-than-human creatures have experiences, have their own reality and their own subjectivity. I agree. What I question, however, is what, more precisely, it is that allows *Nagel* to claim this, given the method and the criteria he subscribes to.

IV

To appreciate what is at stake for Nagel's position here, we need to look more closely at what he says about matter. Like so many of his colleagues these days, when the question about the nature of matter comes up he refers to Bertrand Russell's *The Analysis of Matter*, a work he considers to expound Russell's version of neutral monism, the position Nagel himself endorses early on in *Mind and Cosmos* (2012: 5).

Whereas there is no mention of Russell in that book, Nagel grants him plenty of attention in "The Psychophysical Nexus". Russell famously professed agnosticism about the intrinsic properties of matter: We have absolutely no (positive, verifiable, testable) knowledge of the intrinsic properties of matter which underwrite their causal relations (see Seager 2006: 135). This at least is how Russell's often-cited statement is commonly interpreted: "The only legitimate attitude about the physical world seems to be one of complete agnosticism as regards all but its mathematical properties" (Russell 1927: 270). This means that all physical theory can do is to describe such items as electrons, protons, forces etc. in exclusively extrinsic terms. Physics has no way of telling us what electrons *are*: about their intrinsic nature. It can only tell us how they interact and interrelate with protons, forces and the like. Electrons, therefore, are only accessible for us via their doings with other physical phenomena, similarly defined, and in terms of their impact on our measuring instruments. As for what they are, in themselves, physics is silent.

According to Sam Coleman, if Russell is right concerning physics and its limits, then "physical theory is precisely crying out for intrinsic *somethings* to serve as the doers of the things that it records and relates". Coleman contends that panpsychism satisfies such a need, holding that "the ultimate physical particles are to be thought of, in respect of their intrinsic natures, as loci of consciousness" (Coleman 2009: 88). William Seager, another contemporary philosopher who discusses Russell's agnosticism, follows suit in arguing that "since it is evident that certain configurations of matter generate or constitute conscious states, the intrinsic properties of matter must encompass this power" (Seager 2006: 136).

Must it really? Seager and Coleman are among those who invoke Russell's agnosticism concerning the knowability of matter in an effort to make the case for panpsychism. The intrinsic nature of matter that physical science according to Russell cannot help us grasp is said to be consciousness. Consciousness itself, writes Seager, "provides perhaps the best argument that there are intrinsic properties, and this is exactly why consciousness appears as so troubling or even alien to the scientific picture of the world which deals exclusively with relational or structural features, leaving aside any attempt to grapple with intrinsic natures" (Seager 2006: 143f.).

I find this way of arguing the case for panpsychism less than compelling and assume that Nagel would concur. It fails because it overstretches Russell's merely negative point about the epistemological limits of the physical sciences as Russell was able to conceive them in his time. Seager's attempt at a positive argument takes as its premise that "we are aware of but one intrinsic property of things, and that is consciousness", crediting Descartes (seldom an object of praise in panpsychist circles) with being "the first person" to recognize that "states of consciousness are intrinsic properties" (Seager 2006: 136). Proceeding from this premise, Seager advances the following argument: "It is plausible to assert physicalism – we are physical beings and our consciousness is a feature of certain physical structures. Therefore, consciousness is an intrinsic property of matter" (ibid.). Again, this does not strike me as compelling. What, more precisely, are the "things" of which consciousness is said to be an intrinsic property? "Things" sound very general – do they in this case include nonhuman entities as well as human ones? If both, how to account for the differences between them – not only between humans and nonhuman animals (allowing that they both possess consciousness) but between humans and such things as rocks, or trees, and between the latter and nonhuman animals? And what about such an ontological category as artifacts – is consciousness an intrinsic property in their case too?

Seager seems to give the game away by declaring, in a footnote, that "it is a virtue of panpsychism that it permits us to be physical beings in the face of the difficulties of conceiving of consciousness as a physical phenomenon" (2006: 136f., footnote 5). Granted we are physical beings, and granted we are conscious, the two properties or dimensions co-exist. But again, ontologically speaking that is just one case of a myriad of different ones, and a very limited one at that. By Seager's own admission, "to show panpsychism we need to show that consciousness is both fundamental and ubiquitous" (2006: 137). However, to take ourselves as the model, in effect as paradigmatic for the entire ontological field whose variety of different entities I listed (not exhaustively) above, is question-begging; it leaves the ontological richness and particularity of the various nonhuman entities in the dark, especially with regard to their possible role as loci of experience or of consciousness (not the same thing, as we will see below).

The point I make against Seager has a Nagelian flavor that goes back to his programmatic aim in *The View from Nowhere*: to work out a "general" concept of mind as distinguished from "an anthropocentric concept which conceives all minds on analogy with our own" and regards ourselves as the central instances. What Nagel is after is to "think of mind, like matter, as a general feature of the world" (1986: 18, 19). In what way, if any, does this differ from panpsychism understood as the position that all the elements of the physical world are also mental?

On an abstract level, I see no big difference. In *The View from Nowhere*, Nagel is quick to point out that any attempt to generalize the concept of mind "takes us away from the conditions that make the concept meaningful", in particular the general condition of publicity that "must be met by all concepts". This of course is the Wittgensteinian view that questions any extension of the application of mental concepts from my own case – questions their extension not only to radical others such as rocks or various inanimate entities ("dead matter"), but even to fellow humans. Whether early or late, Nagel never fails to invoke Wittgenstein's so-called verificationalist model when discussing mind. If I am not mistaken, however, there is a growing dissatisfaction with Wittgenstein at this point. Nagel maintains that we can indeed "use the general concepts of experience and mind to speculate about forms of conscious life whose external signs we cannot confidently identify" (1986: 24; 2012: 37f.).

It may seem that we have strayed far from the issue raised above when I brought up Russell's view on matter. Is it not a very different issue from the present one concerning the extension of mind, or the application of mental concepts (Wittgenstein), to other creatures than ourselves?

I shall argue that it is not, and that seeing why is crucial to appreciating the limits to Nagel's assessment of panpsychism.

V

What I find striking is this: Whereas Wittgenstein theorized about mind and Russell about matter, they both conducted their theorizing in the same manner. And even though their respective inquiries tended not to yield positive results but instead agnostic, inconclusive or even skeptical results – amounting to professing that we simply cannot know (the true extension of mind in Wittgenstein's case and the intrinsic properties of matter in Russell's) – neither of them took this to indicate that their specific way of raising the issue may have been wanting, or downright wrong.

What characterizes the like manner that I refer to? Simply put, the model employed in both cases is one of observation: behavior in its capacity as public in Wittgenstein's analysis of mind, electrons, protons, and forces as operative in the doings that physical science records as so many effects that qualify as externally observable (verifiable, repeatable etc.) data in Russell's analysis of matter.

There is no need to reiterate Wittgenstein's model. As for Russell, Nagel quotes his statement in *The Analysis of Matter* where he states that, "we cannot perceive a light-wave", thus frustrating our ability to fully know the latter's nature. In the same passage, however, and also quoted by Nagel, Russell makes the general claim that "to assert that the material *must* be very different from percepts is to assume that we know a great deal more than we do in fact know of the intrinsic character of physical events" (Russell 1927: 263, 264; Nagel 2002: 209).

In later works, however, not cited by Nagel, Russell makes the point that I am getting at even more strongly: "Physics is mathematical not because we know so much about the physical world but because *we know so little*; it is only its mathematical properties that we can discover. For the rest, our knowledge is negative. The physical world is only known as regards certain abstract features of its space-time structure" (Russell 1948: 240). In his paper "Mind and Matter", Russell gives his overall view as follows: "We know nothing about the intrinsic quality of physical events except when these are mental events that we directly experience" (1956: 153).

The model of observation that I have in mind is aptly attested to in these quotes from Russell. By Russell's own admission, employing this model will never actually "show whether the physical world is, or is not, different in intrinsic character from the world of mind" (1948: 240), the latter world, of mind, also being difficult to know as far as its intrinsic properties and its extension (other human minds? nonhuman minds?) are concerned, according to Wittgenstein's adoption of the same model.

What is the point I am making with regard to Nagel? Is it that he can be seen to subscribe to the same model, that of observation, employed by Russell in the case of matter and Wittgenstein in the case of mind?

I am not suggesting that Nagel, in his search for an alternative to "the materialist neo-Darwinian conception of nature" (the subtitle to *Mind and Cosmos*), adopts this model, let alone advocates it as the most fruitful one. Yet he continues to relate his discussion to it and hence to the limits to our knowledge about *both* matter and mind upon which two of its most influential practitioners, Russell and Wittgenstein, insist.

There is something peculiar here. Take Russell. When he states flatly that physics is unable to help as gain knowledge about the physical world, does he really think this has to be so? Or does he find it proof of complete failure on the part of the science in question? Can we conclude from the failure in question that we *have to* fail in our ambition to know about the physical world and its intrinsic properties, with special regard to their difference or similarity to those found in the world of mind? In short, in proclaiming failure on behalf of *one* way of conceiving and doing physics, why didn't Russell immediately go on to propose, and go about employing, *other ones*? Bluntly put, if physics cannot secure us knowledge about the physical world and its intrinsic properties, and in fact admits as much, what do we need physics for?

I do not know how Russell would have responded to these questions, or if he would find them pertinent. Perhaps he thought that physics would have to learn to be modest, to come to terms with the limits to its endeavor that he so bluntly pointed out, and to resist as hubris any aspiration to know it all.

As Galen Strawson has pointed out, in saying that "physics is mathematical not because we know so much about the physical world but because we know so little", by the "physical world", Russell means the nonmental, non-experiential world, "using 'mental' and 'physical' conventionally as opposed terms" (Strawson 2006: 10). Russell would perhaps have been provoked to search for a less agnostic opinion on the subject of matter had he engaged with the following statement by his lesser-known philosopher colleague Arthur Eddington:

> The stuff of the world is mind-stuff, something whose nature is not altogether foreign to the feelings in our consciousness. [. . .] The mental activity of the part of the world constituting ourselves occasions no surprise; it is known to us by direct self-knowledge [cf. Descartes' cogito – AJV], and we do not explain it away as something other than we know it to be – or, rather, it knows itself to be. It is the physical aspects (i.e. non-mental aspects) of the world that we have to explain.
>
> (Eddington 1928: 276f.)

I cannot pursue the specifics of Eddington's argument here. What is important about it in the present context is to do with the suggestion made about the burden of proof: there is an argumentaive asymmetry between explaining and accounting for the experiential and explaining and accounting for the (assumedly) non-experiential in that we know more about – know something with greater certainty about – the experiential than the non-experiential. In short, *to argue for the existence of something non-experiential requires more in terms of argument and proof than does arguing for the existence of experience*. Strawson has his own take on this, turning what amounts to agnosticism in Russell into something of strong argumentative import: "In fact", says Strawson, "we really do not know enough to say that there is any non-mental being" (Strawson quoted in Basile 2009: 184).

On my view, this does not settle the question about *what entities* in the world – here meaning the nonhuman world, entities in addition to human subjects – possess and exhibit experience; that is still left in the dark. But it does say something about where the onus of proof is to be placed: on those doubting or denying the existence of experience, not on those assuming it. But again, that does not take us very far – only as far as allowed for on the very high level of generality on which Strawson, like Russell before him, makes his statements.

Again, it is regrettable that Nagel does not discuss this notion of where the burden of argument lies (or the main burden, to be more precise). We saw that in *Mind and Cosmos* Nagel dismisses panpsychism on the grounds that it "does not provide a new, more basic resting place in the search for intelligibility – a set of basic principles from which more complex results can be seen to follow" (2012: 62). This is a very general criticism. It does not say much about the specifics of panpsychism, let alone what particular version Nagel has in mind. But it does say something about expectations: what a successful theory should be able to do in terms of explanation. Specifically, it should render intelligible "the character and cosmological history of the nonliving material world" (2012: 61), besides accounting for why consciousness "as a physically irreducible feature of certain organisms [. . .], starting from inorganic matter, [is] at all likely" (2012: 65).

These are indeed the sort of questions commonly raised in discussions of panpsychism, the upshot in the large majority of cases being that panpsychism falls short and is duly dismissed. The burden of proof is such that the position comes up short. Discouraged, one looks for other alternatives.

But perhaps the dismissal is too hasty, the negative verdict premature. Take a second look at the questions Nagel asks, being the usual ones. He refers to "the nonliving material world" and to "inorganic matter". He then *contrasts* these entities with the properties associated with the mental, with consciousness and mind, asking how the latter can possibly have originated in, or be showed to be present in, the former, only to answer that they cannot: we cannot know that they are in the manner panpsychism asserts.

I am not suggesting that Nagel is giving an all-out rejection of panpsychism and everything it holds about the natural order. He is not. But he is skeptical, and he tends to conclude negatively, when it comes to panpsychisms' ability to answer the questions he raises, and that he would also raise, we may take it, with respect to other positions. He does not treat panpsychism unfairly or in a selective fashion.

Now take a closer look. In order to establish that panpsychism – taken in its general claim that all the elements of the physical world are also mental – fails to convince in postulating consciousness (or something mental) in "the nonliving material world" and "inorganic matter", we would have to know what the latter are: we would have to know that "the nonliving material world" and "inorganic matter", ontologically speaking, are such as to be nonmental, as to exclude the mental (co) dimension that panpsychism postulates as inseparable from them. But such positive knowledge about what Russell called the physical world, implicitly drawn upon in the way Nagel contrasts the nonliving and the inorganic with consciousness in his formulations, is *precisely what we do not have*, according to the Russellian agnosticism that Nagel himself adheres to, at least in other parts of his discussion of panpsychism.

VI

To the surprise of many of his colleagues in analytic philosophy, Nagel holds natural teleology to be a serious contender when it comes to explaining the big picture: nature and everything it comprises. Equally provocatively, he advocates a form of realism – incorporating value realism, moral realism – that flies in the face of the Hume-inspired subjectivism that to this day dominates Western thought and culture by drawing a strict ontological (as well as logical) line between matters of fact and matters of value, between what is and what ought to be, warning against the "naturalistic fallacy" of inferring from the former to the latter.

In turning to these aspects of Nagel's book, I shall confine myself to the points that are most relevant to the critique I just formulated.

Natural teleology in Nagel's sense holds that the universe is "rationally governed in more than one way – not only through the universal quantitative laws of physics that underlie efficient causality but also through the principles which imply that things happen because they are on a path that leads toward certain outcomes – notably, the existence of living, and ultimately of conscious, organisms" (2012: 67). For Nagel, then, the big question about the natural order immediately takes the form of the *restricted* question about how that order must be explained to do justice to the distinct nature of *one* category of entities within it, namely conscious organisms. The distinctness of such organisms, Nagel implies, is highlighted in humans, understood as conscious organisms with the capacity of reason. What we want to know, then, is how such organisms – ourselves – are possible within a larger theory about the natural order. "To explain", writes Nagel, "not merely the possibility but the actuality of rational beings, the world must have properties that make their appearance not a complete accident: in some way the likelihood must have been latent in the nature of things" (86). The teleology Nagel "wants to consider" (in his preferred way of putting it) is one that seeks to do without the hypothesis of intentional design; it is a teleology that is not committed to an intentional (theistic) explanation of the existence of reason. Unlike all the basic scientific laws discovered so far, the "teleological laws" that Nagel has in mind are temporally historical in their operation; they are "laws of the self-organization of matter, essentially – or of whatever is more basic than matter" (93). These organizational principles must be seen as an "irreducible part of the natural order, and not the result of intentional or purposive influence by anyone", thus avoiding the presupposition of an (intentional, intelligent) designer. Nagel confesses not being confident that what is an "Aristotelian idea of teleology without intention" makes sense, yet declares that he does not see why it does not make sense (93).

At this point in his argument, Nagel would have done well to consult Hans Jonas' *The Phenomenon of Life* and the distinction made there between teleology as a causal mode of nature itself, designated "immanent teleology", and "transcendent teleology" such as "exercised by the creator of the existing system of nature in once creating it as it is" (Jonas 2001 [1966]: 34). Since Nagel explicitly dismisses the latter, let's concentrate on the former, being an unmistakably Aristotelian teleology in that it postulates final causes as the *modus operandi* of and in nature itself, by which is meant the whole of living (organic) nature, not primarily, let alone exclusively, "conscious organisms with the capacity to reason", i.e., human beings.

Significantly, there is no discussion of final causes in Nagel's treatment of teleology in *Mind and Cosmos*. The closest he comes to such a notion when describing what he means by "natural teleology" is when he talks about "things happen[ing] because they are on a path that leads toward certain outcomes" (67), and when he says that "the tendency for life to form may be a basic feature of the natural order, not explained by the nonteleological laws of physics and chemistry" (124).

It is not surprising that Nagel does not stick his head out to endorse final causes as conceived in Aristotelian immanent teleology. As Jonas reminds us, the all-out rejection of final causes was part of the great historical struggle with Aristotelianism which marked the birth of modern science in the sixteenth century, being closely connected with an attack on "substantial forms" and resulting in a proudly anti-teleological mechanistic paradigm where only the category of *causa efficiens* was retained and considered truly scientific, ruling out Aristotle's other three causes, the *materialis* and the *formalis* in addition to the *finalis*.

Jonas' comment on how this rejection came about, and how it has persisted up to the present, is worth citing:

> Regarding final causes, we must observe that their rejection is a methodological principle guiding inquiry rather than a statement of ascertained fact issuing from inquiry. [. . .] The mere search for them was quite suddenly, with the inauguration of modern science, held to be at variance with the scientific attitude, deflecting the searcher from the quest for true causes. It is only then, in the exercise of that attitude, that the negative record is accumulated, viz., by the success of doing without final causes. To repeat, the exclusion of teleology is not an inductive result but an a priori prohibition of modern science. This it can only be it teleology contradicts the very type of being presupposed in possible objects of natural science and thus also the concept of cause proper to such objects.
>
> (Jonas 2001: 34)

In fact, Nagel echoes Jonas' comment when stating that "in spite of the exclusion of teleology from contemporary science, it certainly shouldn't be ruled out a priori" (66). Yet Nagel fails to engage more deeply with the issue, that is to say, with *the primacy of method over subject matter*, with how a certain set of criteria-cum-requirements for knowability have come to determine and delimit what may count as real, so that (positive, verified) knowability becomes tantamount to reality as such, conceptualized by Charles Taylor as the modern scientific primacy of epistemology over ontology mentioned earlier.

In refraining from pursuing the reasons for the rejection of final causes, as well as its rich implications, Nagel's treatment of teleology also omits a related topic, that of anthropomorphism. As Jonas stresses in his historical account, "our very proneness to final explanation makes it suspect"; that's why Robert Boyle complained with disgust that "Men are taught and wont to attribute stupendous unaccountable effects . . . to a certain being . . . which they call nature" (quoted in Midgley 2014: 119); and that's also why Francis Bacon in his *Novum Organum* lists it among the "idols of the tribe", the prejudices innate in human nature, thereby anticipating

Hume's hugely influential psychological take – psychologization, if you will – on the phenomenon of causality. Already with Bacon, that is, the philosophical issue itself – "the inhospitability of nature to final causes – is taken as settled and in no need of argument" (2001: 35). The assertion that so rapidly and so persistently would become an axiom beyond questioning, namely that "final causes have relation to the nature of man rather than to the nature of the universe", carries the all-important implication that "no inference must be made from the former to the latter, which again implies a basic difference between the two" (ibid.). Exterior reality having become entirely detached from the interior reality of thought, *res extensa* from *res cogitans*, the stage was set for the ascription of epistemological monopoly to the perceptual mode of cognition, so that "objectivity" becomes essentially the elaboration of exterior sense data on the lines of their extensional properties, i.e., quantitative and commensurable ones such as size, weight, etc. The upshot of the dominance of "distancing" and objectifying perception, in a series of (in principle verifiable, repeatable) observations of distinct "objects" and sequences of Y following from X (linear causality), was the establishment of "a severe ban on any transference of features of internal experience into the interpretations of the external world". Hence, Jonas maintains, "anthropomorphism at all events, and even zoomorphism in general, became scientific high treason" (ibid.). As far as "final cause" is concerned, then, the nail in its coffin consists in pointing out that it is anthropomorphic. In other words, what fails to qualify as part of the scientific stance to nature, to reality, need not be repudiated by further arguments.

I am not saying that Nagel would have reached a different conclusion concerning "natural teleology" had he reflected on the implications of the rejection of final causes to which Jonas is drawing attention. What Jonas does help us recognize, however, is the selectivity with which Nagel comes forward as a critic of axioms of "the materialist neo-Darwinian conception of nature": willing to openly challenge some of them, passing over in silence and so in effect staying loyal to others, the ban against final causes and against anthropomorphism being two cases in point.

VII

To return to *Mind and Cosmos*, one thing is clear: for Nagel the all-important criterion of validity is the theory's success in explaining as non-accidental the existence of conscious, thinking beings, creatures capable of reason.

On Nagel's view, scientific naturalism not only fails to explain the reality of consciousness and cognition. It fails to explain *value* as well, understood as the *reality* of value, a reality Nagel considers "far less transparent than the reality of consciousness, and also less clear than the reality of cognition of reason in general" (98).

The subjectivist position Nagel contrasts with his realist one holds that in declaring that something – say, an action – is morally right or wrong, we express how we subjectively feel about it and respond to it: favorably in the one case, negatively in the other. (This of course recalls Alasdair MacIntyre's discussion of "emotivism" in his *After Virtue*.) Moral judgments stem from our feelings and attitudes and report on them in the given case; they do not pronounce on their objects – say, the action in question – in the world and how they presumably are "in themselves". It follows that value – deeming something good, right or admirable – is of our making, something we may tend to think that we encounter in the world outside us, as part of its fabric even though it in fact originates with us and our capacities as feeling and judging subjects. No subject, no value.

This is my way of defining the subjectivist position on value, not Nagel's. Yet it is wholly compatible with his. What is important here is Nagel's advocacy of realism, where our responses – "what a cruel thing to do" – "reflect the evaluative truth and can be correct or incorrect by reference to it" (98). Our value judgments are correct, then, when our dispositions are "in accord with the actual structure and weight of values in the case at hand" (100). Nagel gives the example of wanting to avoid grievous harm to a sentient creature. It is *true* that we have a reason to avoid such harm; we have what can be called (though Nagel does not use the terms) an objective as opposed to merely subjective reason to do so, a reason that is to be regarded as "one of the kinds of things that can be true in itself" (102).

Nagel is explicit that his concern is to treat realism as a view about truth and truth conditions. Accordingly, on his view "the dispute between realism and subjectivism is not about the contents of the universe. It is a dispute about the order of normative explanation" (102). This strikes me as a regrettable restriction of scope as far as the assessment of realism is concerned. Subjectivism in the wake of Hume may well want to confine itself to epistemology – the issue of truth conditions – insofar as it informs the judgments we make in moral matters. Realism, however, has much wider scope: it is a metaphysical view, concerned with ontology, with the contents of the universe in its totality. Narrowing the scope from metaphysics and ontology to epistemology and morality means missing out on what is most important for understanding the very notion that Nagel sets out to defend: realist value.

Let me explain what is at stake here. Nagel's discussion portrays the disagreement between subjectivists and realists as a disagreement about what judgments about right and wrong refer to and how to decide their validity. I would suggest that more fundamentally the disagreement is about what the world consists of: whether the world (among other things) consists of *value-possessing entities*, entities of such a kind as to be intrinsically valuable, carriers of objective value, so that in failing to recognize them in this capacity, we would fail to see them for what they truly are, and so would miss out on their particular reality, their distinct ontological nature. Subjectivism is the position that there is no such thing (value is solely for-us, *pour-soi*, something we project onto objects out there in themselves devoid of value), realism that there is, in the strong sense that value is part of the fabric of the world, so that in denying value as existing in itself, as *met upon*, we would fail – epistemically, morally – to do justice to a property of the world.

This contrasts with Nagel's assertion that "realism does not add anything to the catalogue of entities or properties that a subjectivist believes to exist in the world" (2012: 103). It does. The value-possessing entities that I just referred to are indeed a category concerning whose existence subjectivists and realists disagree: for subjectivists there are no such entities; there are entities – namely, human subjects – who project value onto various objects (actions, states of affairs) that we take an active interest in, positively or negatively, by way of attraction or aversion, responding with sympathy and approval when the object is experienced as "good" and with antipathy and rejection when it is experienced as "bad". The Humean point is that goodness and badness reflect on us as subjects, not on the objects.

That realism as understood and advocated by Nagel is opposed to the Humean position goes without saying. What we need to recognize is the limited nature of this opposition. It comes to the fore when Nagel states that "Realism is not a metaphysical theory of the ground of moral and evaluative truth. It is a metaphysical position only in the negative sense that it denies that all basic truth is either natural or mathematical. It is metaphysical only if the denial of a metaphysical position like naturalism itself counts as a metaphysical position" (2012: 101).

But why not admit that realism is a metaphysical theory as good as any? Why proceed in the indirect fashion Nagel chooses here, making realism's status as a metaphysical position dependent on how we understand naturalism?

Metaphysical realism is commonly understood as the position that there exists mind-independent reality. Perhaps Nagel's reasoning is that since the existence and – even more to the point – extension of mind is the very issue about which "the materialist neo-Darwinian conception of nature is almost certainly false" (to quote his book's subtitle), it would only cause confusion if he were to declare himself a metaphysical realist. It would remove attention from what Nagel sees as the major issue: the emergence and presence of mind (or consciousness) in the universe, taken as a question about the possible existence of mind-qualities in beings beside humans.

This question brings us back on track. Above I spoke about value-possessing entities. Are there such? Hume for one is skeptical, and his anti-realism is complete: no entity can meaningfully be held to possess intrinsic value. Value in all cases is a matter of projection, taken as the capacity and deep-seated disposition of human subjects to "colour" the world, to impose upon what merely is (what is merely factual and as such neutral, indifferent) some notion about how it ought to be. The imposition of such an ought is the origin of value, how it comes about: always of our making and dictated by our feelings, desires, and beliefs, and never part of the world and the way things are.

Kant took a different view: value-possessing entities exist, but only and exclusively in the case of humans (persons). Value takes the form of *Würde*, meaning dignity in the strong sense of inviolability. It is a constitutive property of humans, defining of our full-blown moral status as ends in ourselves and setting us apart from, and morally above, all other creatures in the world, even animals. Kant's model is symmetric in that it postulates a specific mutuality: it requires the faculty of reason to be able to recognize the intrinsic value of others; and only creatures in possession of the (intellectual) capacities required to enact such recognition deserve recognition. One subject's powers to recognize others as intrinsically valuable are of the same kind as what commands those others to reciprocate the recognition. Everything involved here hinges on the possession of reason. Nonhuman animals, indeed everything nonhuman in the entire natural world, are seen as non-possessors of value, and as non-candidates for full moral standing (inviolability), on the grounds that they lack reason. To lack reason is to lack the capacity for freedom and for initiating action, what Kant calls spontaneity. To lack reason is to lack agency. Only human agents can recognize – in Kant's double sense of epistemically *erkennen* and morally *anerkennen* – entities in the world possessing value, these being limited to beings who are themselves capable of agency. Hence the symmetry.

There is a deeper issue involved here, that of the connection – if any – between possessing mind and possessing value. Can value reside in something devoid of mind? Is the latter a necessary condition for the former? We shall return to these questions below.

VIII

Nagel begins his chapter on value by stating that "the problem of value in the natural world includes but goes beyond the problems of the place of consciousness and of cognition in general" (2012: 97). This sounds promising. He then proceeds: "It is clear that the existence of value and our response to it depend on consciousness and cognition, since so much of what is valuable consists in or involves conscious experience, and the appropriate responses to what is good and bad, right and wrong, depend on the cognitive recognition of the things that give us reasons for and against" (ibid.). Value, Nagel contends, has to do "specifically with the practical domain – the control and assessment of conduct" (ibid.).

Really? Does it take conduct for value to be in play, to be at stake? If so, what qualifies as capable of conduct? And what qualifies as capable of assessing that conduct in terms of value?

In the course of a single sentence, Nagel very significantly narrows the horizon within which he conceives of and discusses value. He moves from asking about the "place of value in the natural world" to pinning value down to the "practical domain", which he then immediately equates with conduct.

There is something unmistakably Kantian about this. Say "conduct", and the ones capable of producing (authoring) it as well as responding to it and assessing it – in terms of value – are presumably human agents and such ones only. Of course, nothing forbids one from attempting to extrapolate from this, so as to suggest, say, that some category of nonhumans are also capable of conduct in the double sense involved, that of bringing it about and that of responding to it. But Nagel fails to explore such a possibility.

The narrowing I see at work in Nagel has him ending up in the vicinity of Kant, who, like Hume and despite their differences, subscribes to a purely anthropocentric scope for value as well as mind. We saw what the symmetric formula amounts to: only beings able to recognize value possess value.

Recall Nagel's formulation: "It is clear that the existence of value and our response to it depend on consciousness and cognition, since so much of what is valuable consists in or involves conscious experience" (2012: 97). Various claims are involved here. Let us grant, for the moment, that *responding* to value requires certain capacities of the sort we see operative in consciousness. The question is whether accepting this commits us to Nagel's initial claim: that the *existence* of value (also) depends on consciousness and cognition. I think not. And I find it noteworthy that Nagel doesn't argue for the claim he is making. It is as if he presumes his readers will readily agree with it and not ask for a separate argument. The tacit acceptance thus assumed is an indication that the position Nagel is subscribing to is a conventional and widely shared one, among his readers and in the culture they (we) belong to.

To see where I differ from Nagel, let me put the symmetric model I have been referring to like this: value is an entity that involves a subject and an object. For lack of someone for whom value is, there is no value: for want of actively being recognized, and so creatures capable of doing that, value simply would not be. As for what value is, Nagel doesn't say much, except that "so much of what is valuable consist in or involves conscious experience" (2012: 97). Hence the symmetry: in describing what it is about the entity said to have value, to be valuable, that makes it so, the qualities focused on are precisely those said to be present in the subject who responds to value (value-entities in the world), namely conscious experience. The symmetry makes for a circle: the subject relates to the object, and the object is described in terms of properties operative in the subject.

The fact is that most of the philosophers who discuss value have virtually nothing to say about what value *is* – Nagel being no exception. Typically, the discussion – and disagreement – concentrates on *where value is to be located*: in humans only, or in nonhumans as well, or in some wider variety of entities. Value is theorized and discussed in terms of location and extension, not in terms of being, of what it is in itself as opposed to what value is for us, framed purely anthropocentrically at that.

Now consider the following argument. All living beings have an indisputable and objective interest in going on doing so, and thus, in each individual case, they embody a stance of non-indifference, non-neutrality, with respect to the difference between life and death, being and not-being: the former is better than, is superior to, the latter. It is so as a matter of fact, and this fact is the objective fact of value: that life, that being, is the valued state. The fact of this value is not imposed from without, is not the product of projection or attribution from some external source or point of view. Instead, it resides in the dimension of being – *in re* – as one of its essential properties.

Qua existing in itself, value calls for response. Protection-intending respect is the appropriate response to the value of the various entities the unfolding world consists of. Value as met upon in outer reality issues a demand on us as human agents to act so as to observe what the protection of that value requires, value being often a precarious quality, a quality whose safeguarding the entity in question may not itself be capable of securing. This demand is normative, it has the form of an ought, only this ought does not originate in me (in consciousness, intellect, volition) but stems from the world my care responds to – the world to which care is *the* appropriate response, factually and morally appropriate, there being no meaningful distinction between the two. In thus responding, I act as addressee, not initiator; being able to respond in this sense is what response-ability is all about: it takes the form of responsibility.

On this realist view, then, value concerns that which resides in specific entities in the world, as opposed to the notion, a long-standing one in Western philosophy, that value is a product of projection onto a value-devoid world on the part of the (human) subject – the very notion that informs the naturalistic fallacy and that, crucially (yet seldom noted), is maintained in Kant no less than in Hume and Nietzsche. Anthropocentrists from Hume to the present get it exactly wrong when they posit humans as the sole source as well as the locus of value. Value exists perfectly in nonhuman entities, not only in animals but in trees and plants, entities each of which pursues, in their species-specific manner, a good of their own. To speak of value as existing factually and objectively in various entities is not to deny that the *actions* of humans, more than the activities of any other species, may negatively affect the fate of all such real-life instantiations of value, as the entry into the Anthropocene demonstrates with such alarm. But then again, humans' sheer power in being able to undercut, endanger, and perhaps eventually all-out destroy value in the world that we are part of does not mean that we are the creators – origin, source – of the value destroyed. We destroy what is given independently of us, what has historically evolved prior to us, a quality whose true character is that of a gift we should treat with gratitude and awe, since as a species we owe our existence to the flourishing of a host of other species and life forms.

The simple fact is that the entire nonhuman world constitutes the *sine qua non* of humans. This means that any insistence on the part of humans to be exclusive possessors of value will prove self-destructive. Whitehead put it like this almost one hundred years ago: "Any physical object which by its influence deteriorates its environment, commits suicide" (1967 [1925]: 109).

Realism in the version I advocate here holds that an ought is indeed present *in re*, in a piece of reality existing independently of us humans and our workings. Whether the ought thus said to be encountered in the natural order (Nagel) is to be taken in the sense of biocentrism or ecocentrism, or in a more limited one, concentrating on individuals instead of systems, and on some living organisms rather than all, is a big question to which we shall return below. At its most radical, it is a question about the existence of the qualities of mind and value not only in – something, or everything – *living*, but in physical reality in its entirety, spanning even the inanimate and inorganic. To affirm the latter, of course, is precisely what the most radical versions of panpsychism are up to.

Before returning for a final look at Nagel's position, I need to say a bit more about the sort of realism I propagate in holding such entities as eagles and plants to have intrinsic value regardless of my, or any other human's, actually experiencing the entity. What realism denies is precisely the notion that the value of that entity depends on such an act of experience. Even in the case of the "last man dying", the eagle or plant surviving humanity's existence on earth would possess exactly the same value as when co-existing with humankind. In other words, the existence – contingent at that – of *one* species cannot be mandatory for the value of others. Value concerns what *resides in some specific entity in the world*, not in something to do with properties found in, or even more basically: the existence of some *other* entity that happens to relate to, or seek knowledge about, the entity in question.

Joel Kovel puts it like this:

> Intrinsic value is the value given to reality irrespective of what we do to reality. In terms of reality-as-nature, intrinsic value is a kind of ablation of our productive power; that is, we intrinsically value the nature that we have done nothing to, that will always stand and beckon, that is our primordium and cosmos – not for sale, and not to be made into a commodity, rather, the "suchness" of nature, its intrinsic being, both sensuously immediate and eternally beyond our ken and grasp. It is the sense of the world conveyed in words like

> "wonder", "awe", or simply the quiet appreciation of the day without respect for what can be made out of it – including, of course, the making of money.
> (Kovel 2002: 212f.)

IX

The value realism I have sketched holds that value exists objectively in a host of nonhuman entities and independently of humans. The distinctness of nonhuman entities is just as important to account for as that of humans. In much of his discussion, however, Nagel's interest in value realism concentrates on "the distinctive conception of human beings" that is implied by it, allowing us to recognize the way human agents can be motivated by their apprehension of values and reasons. As indicated earlier, this is a take on value realism that immediately assumes the form of questions like "Why did (person) A choose to help (person) B?", and answering "because he saw she was in pain". Nagel's realist point is that pointing to the existence of pain is pointing to an objective fact in the world; pointing to that fact is a *reason* – a sufficient reason, a rational reason – to explain one's action and what motivated it.

Within its proper context, that of morality, this is fine. But value realism, not to mention metaphysical realism, has a much wider scope than that of explaining what motivates humans in their moral conduct toward fellow humans. Is this wider scope not precisely what one would expect Nagel to concentrate on as part of his proclaimed endorsement of natural teleology?

Indeed, Nagel observes that on his understanding of natural teleology, the natural world would have "a propensity to give rise to beings of the kind that have a good – beings for which things can be good or bad". Importantly, he adds: "These are all the actual and possible forms of life" (2012: 121). Good or bad, then, are values to be found "in the lives of other conscious creatures, and perhaps even in forms of life devoid of conscious experience" (118). This certainly sounds expansive, acknowledging value in a much wider context than the purely anthropocentric one that we just saw Nagel focus on in explaining the objectivity of reasons for moral conduct. What Nagel is committed to in this new move is an asymmetry of sorts, one contrasting sharply with the Kantian model of symmetry I mentioned above. "Only beings capable of practical reason can recognize value", Nagel writes, "but once they recognize it, they find it in the lives of creatures without practical reason" (118).

This marks a crucial break with Kant. Kant's insistence that only recognizers of value themselves truly possess value, as brought out in his linking person and dignity (*Würde*) by way of reason, rendering all nonhuman creatures mere means to human ends, is being rejected in favor of an acknowledgment of value in entities incapable of (practical) reason. In taking this step, Nagel not only contradicts Kant, he also contradicts his earlier statement that value specifically has to do with the practical domain – the control and assessment of conduct (97).

What then does Nagel's more expansive position on value amount to? He puts it like this: "In the broadest sense [value] is probably coextensive with life, though how much of this value we humans have any reason to care about is a question I will leave open. It seems too simple to hold that only the value in conscious lives generates reasons. As [Thomas] Scanlon says, it would be callous and objectionable to cut down a great tree just for the fun of trying out one's new chain saw" (118). As for the latter, one would have expected something stronger: Kant would join Nagel and Scanlon in finding the tree-cutter's action objectionable, though for other reasons: agent-related ones rather than tree-related ones.

Be that as it may, the key issue has not to do with moral conduct but with *the place of value in the natural order*. Nagel explicitly acknowledges "the appearance of value in the myriad forms it takes in the variety of lives capable of having a good" (118). That nonhuman life forms and species pursue what is good for them, proper for them, in accordance with their particularity as *this* species rather than that, is of course in keeping with natural teleology. Beyond that, however, it is a notion that invites comparison with the ways in which recent philosophers of the environment have sought to establish the inherent value of nonhuman species and even entire ecosystems – the biocentrism of Paul Taylor (1986), Holmes Rolston (1988), J. Baird Callicott (1989) and the deep ecology of Arne Næss (1989) spring to mind. In *Mind and Cosmos*, however, like in his other books, Nagel pays no attention to these contributions.

Does it matter? I think so. In abstaining from relating his natural teleology in general and his value realism in particular to ecophilosophical debates concerning these topics, Nagel leaves his readers in the dark as to the more precise implications of his position. Crucial philosophical questions are buried here: What is the proper locus of value? The individual or the whole? The token or the type? Granted that (intrinsic, non-instrumental) value resides in nonhuman life forms – should value be confined to entities manifesting life, or should value be seen as residing in nonliving entities as well, say, in rocks? Is value coextensive with physical reality, so that everything factually existing, simply in virtue of existing, manifests value? Or would such a radical extension empty value of its meaning? If a division must be made between things possessing value in the strong sense that value is inseparable from its ontological makeup, from the being it is, and things not possessing value, where can you draw the line, and with reference to what criteria?

The theory of value worked out by Holmes Rolston strikes me as particularly instructive in pondering these questions. His own position is that of a biologically informed holism in which three types of value – commonly separated and seen as conflicting or mutually exclusive – are argued to intersect and interact, each one depending on and helping sustain the others, namely intrinsic, instrumental, and systemic value. Rolston offers a host of biologically based counterexamples to the focus on individuals – "typically sentient and usually persons" – that he finds characteristic of Western thinking (Rolston 1988: 143). It is not only that anthropocentric notions of value must be dismissed; so too must individualistic ones of all stripes. The reason is to do with a simple yet profound fact: "In an evolutionary ecosystem it is not mere individuality that counts; the species is also significant because it is a dynamic life-form maintained over time by an informed genetic flow. The individual represents (re-presents) a species in each new generation. It is a token of a type, and the type is more important than the token" (ibid.).

Extinction (an all-too-frequent reality these days that I shall discuss more fully in later chapters) is given particular attention by Rolston in that it concerns species, not individuals: "The main thing wrong is that extinction shuts down the generative processes" (ibid.). Every extinction is "an incremental decay in this stopping of the flow of life" and hence "a kind of superkilling" in that it kills forms beyond individuals, essences beyond existences; it kills collectively, not just distributively. "To kill a species is to shut down a unique story" (1988: 144, 145), irreversibly so.

Rolston recognizes that species are not moral agents. Nevertheless a biological identity is actively defended, sustained, and protected in the species, which is thereby committed to value, to the value of going on existing. Now, because a species lacks moral agency, reflective self-awareness, sentience, or organic individuality, Western thought has it that they cannot count morally: to lack such (intellectually oriented) capacities is tantamount to lacking moral standing. Against this long-standing view, still dominant in our culture, Rolston points to simple biological facts, facts whose implications are precisely value-laden and thus normative, entailing an ought, an injunction built into the facts and inseparable from them as to how things should be, whereby disallowing them to be such involves a moral violation against them, a violation for which moral agents – yes, we humans, we individuals – can be held responsible.

This is Rolston's argument:

> Each ongoing species defends a form of life – on the whole, good things; pro-life impulses that have achieved all the planetary richness of life. All ethicists say that in *Homo sapiens* one species has appeared that not only exists but ought to exist. But why say this exclusively of a latecoming, highly developed form? Why not extend this duty more broadly to the other species (though not with equal intensity over them all, in view of varied levels of development)? These kinds too defend their forms of life. We humans are the product of such defenses during long eons past. Only the human species contains moral agents, but perhaps conscience *ought not* be used to exempt every other form of life from consideration, with the resulting paradox that the sole moral species acts only in its collective self-interest towards all the rest
>
> (Rolston 1988: 143f.)

How does this square with Nagel's case for natural teleology? On Nagel's view, a teleological explanation becomes plausible once we recognize that "an explanation of the appearance and development of life must at the same time be an explanation of the appearance and development of value" (2012: 121). What exactly does this mean? Are life and value parallel dimensions of one and the same evolutionary process? Are they inextricably intertwined, inseparable aspects of the same process?

Nagel states that "what explains the appearance of life is in part the fact that life is a necessary condition of the instantiation of value, and ultimately of its recognition" (ibid.). Recognition by whom? Nagel does not say; that is, he does not raise the question in the context of the parallelization of life and value that he propagates in the passage quoted. The closest he comes to concretizing what he means by recognition is the observation that "If we were not inclined to recognize objective reasons for action, and were motivated exclusively by our desires, we would have no reason to believe in the existence of value in a realist sense" (2012: 122). This is no doubt consistent with his earlier criticisms of subjectivist notions of value, yet – again – it is one that narrows the wide cosmological scope that a natural teleology worthy of the name aspires to account for. It is fine that Nagel defends objective reasons for action over against merely subjective ones. What limits the reach of his argument, however, is that it so clearly applies to moral conduct only, and so to the kind of agents capable of such – humans only.

A couple of statements by Rolston may help us appreciate the significance of this narrowing of perspective in the course of Nagel's defense of a natural teleology:

> Things do not have their separate natures merely in and for themselves, but they face outward and co-fit into broader natures. Value-in-itself is smeared out to become value-in-togetherness. Value seeps out into the system, and we lose our capacity to identify the individual as the sole locus of value. [. . .] Intrinsic value, the value of an individual "for what it is in itself", becomes problematic in a holistic web. True, the system projects such values more and more with its evolution of individuality and freedom. Yet to decouple this from the biotic, communal system is to make value too internal and elementary, to forget relatedness and externality. The humus and the brooklet are valuable (able for value) and are of value (carry values objectively) because in that matrix the trillium springs up. [. . .] Concern about populations, species, gene pools, habitats requires a corporate sense where value can also mean "good in community". Every intrinsic value has leading and trailing *ands* pointing to value from which it comes and toward which it moves. Natural fitness and positioning make individualistic fitness too system-independent. Intrinsic value is a part in a whole, not to be fragmented by valuing it in isolation.
>
> (Rolston 1988: 216, 217)

Nagel and Rolston are both declared realists, defenders of value realism and of a notion of objective value where value is seen, as met upon in the things themselves and as part of their makeup. Despite this agreement, however, they differ in important respects. When Rolson says "valuing", what he refers to is an always-ongoing *activity* within the value-possessing entities, all of them at that; an activity, moreover, that unfolds in the relations the myriad of entities are part of in the strong sense of being constituted, in their distinct particularity, by virtue of partaking of those relations with "others". This view recalls the relational ontology propounded by Hegel; an ontology that, liberated from the framework of philosophical idealism, informs ecology as a science, inaugurated by Ernst Haeckel in 1866 (see Callicott 1989: 105). In stark opposition to the atomist ontology of Hobbes, part and parcel of his mechanistic world view, relational ontology insists that no entity is self-sufficient, that each thing becomes what it is and sustains itself in its distinctness from others by way of relating to them: being exposed to others is not a threat to the entity's existence but its *sine qua non*. This is the gist of Whitehead's observation, cited above, that "any physical object which by its influence deteriorates its environment, commits suicide" (1967: 109).

Rolston's biocentric version of what I call relational ontology has some consequences for his understanding of value that may help us appreciate exactly where he and Nagel differ. On Rolston's view, each (living) entity possesses intrinsic value (in addition to instrumental and systemic value). Yet the entity's doing so is not of its own making: it is in virtue of the contributions from the entity's myriad others that it comes into being and sustains itself as the being it is, enabled to pursue its species-specific good by way of exchange with the other ones that likewise are busy pursuing theirs. Contra Hobbes, existence is not a matter of competition and struggle but of exchange and mutual dependency; it is not a zero-sum game but a joint and truly communal undertaking. The value that exists in the one (individual, exemplar, token) does so thanks to the many. True, we typically concentrate on, and seek to protect, the value of the individual tiger. But what we thereby value, recognize as value-able, can only exist on the condition that the extra-individual whole of which the individual is a part goes on existing: the tiger-in-the jungle. To sum up, the value manifest *in* the individual is the result of the value-production taking place in the system as a whole.

What about humans? As for the production, the bringing-forth of value, of things existing that pursue what is distinctly good for them, the role of humans is neither constitutive nor necessary. We are a latecomer to the whole value-producing evolutionary process that has been busy producing and sustaining value in the myriads of things, in the long line of generations that have come and gone. Value-production as operative on planet earth is something that has been perfectly able to take place, to unfold in the direction of ever-growing complexity and richness, without any role on our part, predating our emergence on earth as relative latecomer, thus inviting modesty, the opposite of hubris. Value-production did and does perfectly well without us.

Value, then, is precisely not in the eye of the (human) beholder.

2 Panpsychism as "inner physics"

Whitehead's project

In his book *Unsnarling the World-Knot: Consciousness, Freedom, and the Mind-Body Problem*, David Ray Griffin offers what Thomas Nagel stops short of: a full-fledged exposition of panpsychism. In doing so, Griffin discusses the most important contributions to the philosophy of mind over the last decades, counting, besides Nagel, John Searle, Daniel Dennett, Jaegwon Kim, Colin McGinn, and Galen Strawson.

For my purposes, the importance of Griffin's discussion is mainly to do with his defense of what he calls *panexperientialism*, a distinct form of panpsychism that, in Griffin's elaboration, is strongly influenced by the philosophy of Alfred North Whitehead. We will need to dwell on Whitehead in order to grasp what I found Nagel to be short on: the ontological as distinct from epistemological implications of a panpsychist doctrine that aims to overcome the legacy (one is tempted to say the strange non-death) of Cartesian dualism, not only within academic philosophy but in modern culture.

I

Non-philosophers will probably be surprised to learn how influential the Cartesian way of thinking about the world in general, and mind and consciousness in particular, still is within cutting-edge analytic philosophy. Although the answers sought may not be Cartesian – in fact they will often be declaredly not so – the way in which the basic questions are asked, and possible solutions assessed, continues to be conceptually as well as substantially shaped by Cartesian dualism. To mention a few of the examples given by Griffin, there is Geoffrey Madell's admission that "the nature of the causal connection between the mental and the physical [. . .] is utterly mysterious"; Karl Popper's and John Eccles' conclusion that understanding how interaction occurs between nonphysical mind and physical brain is perhaps "impossible"; Nagel's statement that the attempt to develop a physicalist account of mind has led to "extremely implausible positions"; and Colin McGinn's contention that "our present perplexity is terminal", so that we will "*never* be able to resolve the mystery of how consciousness could emerge from the brain" (quoted in Griffin 2007: 3f.).

Deviating from this common lament that consciousness is the "anomalous thing" (McGinn), there is (as mentioned in our above discussion of Nagel) Galen Strawson's turning of the tables, expressed in his statement that "in fact, we really do not know enough to say that there is any non-mental being", echoing Arthur Eddington's contention in his long-forgotten masterpiece *The Nature of the Physical World* (1928) that "it is the physical aspects (i.e., non-mental aspects) of the world that we have to explain", as opposed to "the mental activity of the part of the world constituting ourselves" that "occasions no surprise", being "known to us by direct self-knowledge" (quoted in Skrbina 2009: 56, 184). Witnessing the terminal disagreements, hesitancies, and turn-arounds over what "matter" is, what "mind" is, and which of them we know the most or the least about, outsiders may be excused for doubting if there is such a thing as progress in philosophy.

John Searle aptly sums up the present deadlock when he says that the basic problem is the acceptance of the Cartesian dualism's insistence that if something is "physical" it cannot also be "mental", and if something is "mental" it cannot also be "physical" (quoted in Griffin 2007: 6). Griffin's own position is that the seemingly endemic failure to solve the so-called mind-body problem is more deeply rooted in the Cartesian understanding of the body than in that of the mind. In other words, and to anticipate a major topic, Descartes may have been right about mind but was probably wrong about matter.

Griffin points out that a key motivation in seventeenth century thinking was to be able to assert the immortality of the soul by declaring its absolute separateness from the body. Even more important, though, was probably the permission gained by declaring matter totally inert and insentient to *use the entire nonhuman world for human purposes with impunity*. In addition to the moral, or rather immoral, consequences of such a view, there is an important point to be made about the primacy of method over subject matter: only what allows for observation and measurement qualifies as real; matter *is* simply what can be treated by the method employed by modern science as developed by Descartes, Isaac Newton, and Robert Boyle (see Shapin and Schaffer 1985). This may seem an old story about views long abandoned. On Griffin's view, however, materialism up to the present involves "the creation of a total worldview out of the method intended by the seventeenth-century thinkers to be used for only a portion of reality" (14). The significance of the primacy of method goes deeper than the tendency to notice only what one is looking for. The primacy also helps distort, deny, and thwart the heterogeneity of the phenomena, the richness of the world, according to the homogenizing, commensurability-enforcing constraints of method, that is to say, of a device developed by the epistemic subject so as to seize upon, manipulate, and put to use an outer world divided into so many objects.

As noted above, a hallmark of Cartesian dualism is the idea of mutual exclusion: what mind is, matter is not; and what matter is, mind is not. Hence, the difference between the two substances (Descartes' term, now out of fashion) could not have been ontologically deeper. Mind has an "inside", filled with subjectively rich mental and experiential content in the form of thoughts, desires, feelings, and volitions, and it is thereby accorded intrinsic value. Following the Jean-Paul Sartre who wrote *Being and Nothingness* (1943), dragging Cartesian metaphysical dualism into the then-novel fields of phenomenology and existentialism, mind is *pour-soi*, for-itself, capable of standing over against and transcending (negating) whatever is "given" in the outer world, hence free from its constraints in any given spatio-temporal setting. Matter by contrast is all "outside", devoid of any capacity for relating-to and engaging-with, of mentality in general and so of purpose, ends, and meaning. In short, it is everything we associate with subjectivity and interiority. A bare *en-soi*, in-itself, matter is simple self-identity and as such static and mute; it is what is and cannot be (become) otherwise. Matter is the principle of inertia incarnate. If matter alters or moves, it does so not owing to an innate ability to initiate any such change but wholly as a result of external impact, whose force dictates the change affected, making it purely reactive. Spatial extension is matter's only essential attribute.

John Searle is one of the leading philosophers today who has sought to question the legacy of the conceptual dualism inherited from Descartes and maintained in twentieth-century Continental philosophers such as Sartre. In doing so, Searle focuses especially on the notion that if something is "physical" it cannot also be "mental" and vice versa. According to Griffin, however, Searle, like many of his present-day colleagues in analytic philosophy, does not go far enough in his criticism. He continues, that is, to "assume that *most* physical things are *not* also mental" (47). Griffin's claim is that it is precisely this assumption that renders the problems of the various dualisms and materialisms insuperable. It is tempting to comment that that is the way it goes: if you separate from the very first what should not have been separated, you will strive forever afterwards to bring them together again.

In line with the common view, Griffin holds that the chief problem of dualism has always been to explain "how two totally different types of things could causally influence each other". In other words, "how could that which is spatially extended and embodies physical energy but is devoid of any duration, therefore of any 'inside', therefore of any feelings and desires, and therefore of any intrinsic value, be capable of exerting causal influence on a nonphysical mind?" (49). The kind of "interaction at which our minds and bodies seem quite good", Griffin comments, is precisely what Cartesian dualism cannot explain, as Descartes himself admitted when he responded to Princess Elisabeth that the human mind is not capable of conceiving of how it is possible for mind and body to interact. Since their doing so is such an indubitable fact to us, not being able to explain it is all the more embarrassing.

But why agree so readily that the most serious challenge for Cartesian dualism is what Descartes himself regarded as such? More importantly for my purposes in what follows, why take the mind-body problem, understood as the fact of interaction between the two *in the case of humans*, as paradigmatic? If we want to question the Cartesian mind-matter dualism, meant to comprise everything that exists, why concentrate on such a special case as that of humans having minds and bodies? What about entities that are clearly included in Descartes' matter, yet that do not exist in the form of bodies? What about rocks? Or other sorts of material entities where there – at least on the face of it – is no interaction problem, i.e., no co-presence of mind or something mind-like to account for, given the Cartesian dualism?

The poverty of the Cartesian notion of matter that we begin to identify can be illustrated by observing that an animal and a rock are placed in the same category: qua instances of matter, they are alike in everything that Descartes took to distinguish matter as opposed to mind: in their passivity and externality, in particular. Sartre – nowhere mentioned by Griffin – retained the same conceptual poverty, but his doing so has never been a central topic in the criticisms leveled at his philosophy. In aspiring, just like Descartes, to lay out the ontological structure of reality in its chief categories, Sartre makes no distinction between a gorilla and a rock, or between a gorilla and a tree. He makes no distinction, that is, within living nature nor between living nature and so-called dead matter, between nonhuman animals and organisms, such as plants and trees, on the one hand and matter taken as sheer inertia, devoid of any interiority, on the other. Just like in Descartes' notion of *res extensa*, the former is collapsed into the latter, with human consciousness posited as the only instance of transcendence of inertia that there is.

Even though Griffin gives primacy to the mind-body problem as it shows up in humans, thereby greatly narrowing the scope of Cartesian dualism in general and its notion of matter in particular, his discussion centers on how the emergence of experience out of nonexperiencing entities is conceivable – a question he is convinced that dualism and materialism alike, however sophisticated their present-day versions, will never be able to answer. Interestingly, Griffin quotes Nagel's employment of Sartre's distinction, to the effect that "one cannot derive a *pour-soi* (for-itself) from an *en-soi* (in-itself)". The gap, says Nagel, is "logically unbridgeable. If a bodiless god wanted to create a conscious being, he could not expect to do it by combining together in organic form a lot of particles with none but physical properties" (Nagel 1979: 188f.; Griffin 2007: 63f.). Nagel's position is that it is inconceivable that a subject, understood as "something that it is like something to be" (to allude to his famous paper "What is it like to be a bat?"), could arise out of mere objects naturally. It is important to recall that in asking the question about the bat, what Nagel is after is to know what it is like *for a bat* to be a bat. But even if we are perfectly willing to grant that a bat does have experience – something that Descartes and Sartre deny – it is "only if we knew what they were like" that we could really understand what that experience is, assuming it to be so different from the experience *we* have as to constitute bats as what Nagel calls "a fundamentally *alien* form of life" (Nagel 1979: 168). So whereas we would make a mistake in dismissing as meaningless the notion that bats have experiences "fully comparable in richness and detail to our own", the concepts that would enable us to understand bat experience "may be permanently denied to us by the limits of our nature" (Nagel 1979: 170). This conclusion seems inevitable given Nagel's premise that "even to form a *conception* of what it is like to be a bat (and *a fortiori*) to know what it is like to be a bat) one must take up the bat's point of view" (Nagel 1979: 172).

This may seem an exceedingly strict premise, an order too tall to allow for the "subjective universals" that Nagel has sought to work out elsewhere, hoping to establish "a general concept of experience that extends far beyond our own or anything like it" and to avoid the pitfalls of "an anthropocentric concept which conceives all minds on analogy with our own" (Nagel 1986: 18, 21). How, then, can you do justice to the fact that we, as a matter of fact, "include the subjectively *un*imaginable mental lives of other species, for example, in our conception of the real world without betraying their subjectivity by means of behaviorist, functionalist or physicalist reduction" (1986: 21)? Are we on solid ground when we recognize as real the mental lives of other species, species we will probably never be able to position ourselves so as to take up the peculiar point of the species in question?

It is worth asking whether Nagel's premise is valid: that the facts of experience, taken as facts about what it is like *for* the experiencing organism, are "accessible only from one point of view" (1979: 172), that of the organism (species) in question. To his credit, Nagel follows through on his premise, even if doing so entails reaching an impasse: "If the subjective character of experience is fully comprehensible only from one point of view, then any shift to greater objectivity – that is, less attachment to a specific viewpoint – does not take us nearer the real nature of the phenomenon: it takes us farther away from it" (1979: 174).

In a manner that I consider symptomatic in contemporary discussions of these issues, we have, probably without noticing, moved all too quickly from the widely shared thought that a *pour-soi* (for-itself, mind) cannot be derived from an *en-soi* (in-itself, matter) to reflections on the seemingly insuperable problem of how *one* instance of *pour-soi* (we humans) may understand *another* (say, bats) in a way that does justice to the distinctness of experience as had from *that* other's point of view. At some point in the reasoning, the *en-soi* slipped out of the picture. If you say matter, philosophers will ponder mind: not in the sense of seeing mind in matter, but in the sense of changing the subject. This is regrettable, since the understanding of matter may well prove to be what is most problematic.

To see where this takes us, consider Griffin's assertion that Searle's attempt to portray "mental properties as fully 'ordinary', just one more example of ordinary physical properties", amounts to what is "a prime example of a *category mistake*" (Griffin 2007: 64). Griffin's argument is worth quoting in full:

> The alleged emergence of subjectivity out of pure objectivity has been said to be analogous to examples of emergence that are different in kind. All of the unproblematic forms of emergence refer to *externalistic features*,

> features of things as *perceived from without*, features of *objects for subjects*. But the alleged emergence of experience is not simply one more example of such emergence. It involves instead the alleged emergence of an "inside" from things that have only outsides. It does not involve the emergence of one more objective property for subjectivity to view, but the alleged emergence of subjectivity itself. Liquidity, solidity, and transparency are properties of things *as experienced through our sensory organs*, hence properties for others. Experience is not what we are for others but what we are *for ourselves*. Experience cannot be listed as one more "property" in a property polyism. It is in a category by itself. To suggest any analogy between experience itself and properties of other things as known though sensory experience is a category mistake of the most egregious kind.
>
> (Griffin 2007: 65)

To identify this persistent category mistake is to bring home the point that Nagel addresses in his talk of "faulty analogies", namely that we have no real analogies for the alleged emergence of experience out of wholly insentient entities. Recognizing this has driven many a contemporary philosopher of mind to despair. Colin McGinn, for example, exclaims that "one is tempted, however reluctantly, to turn to divine assistance: for only a kind of miracle could produce *this* from *that*. It would take a supernatural magician to extract consciousness from matter, even living matter" (McGinn quoted in Griffin 2007: 67). McGinn's conclusion is as frank as it is damning for a whole paradigm: "Consciousness appears to introduce a sharp break in the natural order – a point at which scientific naturalism runs out of steam" (ibid.).

The question is whether this impasse is inevitable. It certainly appears to be given the premises of Cartesian dualism, making it all the more important to finally leave it behind. What we must do, then, is seek an understanding of the natural order that accommodates consciousness – or more generally, experience – as an inseparable dimension of nature rather than somehow – mysteriously – added to it. We must, Griffin insists, overcome the widespread assumption that we "know" that conscious experience has somehow emerged out of insentient matter, an assumption that has made it impossible to get to the bottom of how such emergence is possible. Whereas we are clearly right about *what* emerges – consciousness and experience being facts we cannot, on pain of performative contradiction, deny – we may have been wrong about what those phenomena emerge *from*. If the ontological abyss between the two is rejected, the problem of emergence may be solved. If, that is, the ultimate elements of nature can be shown to be of such a kind as to be *conducive* to experience, or even to *inhabit* experience (in some form and degree), instead of excluding it and being alien to it – mind as not-matter, matter as not-mind – then hopefully progress can be made.

II

What is the most promising philosophical alternative to the Cartesian intuition about matter? Griffin's answer is panpsychism, a doctrine never taken seriously by scientific naturalism, materialism, and physicalism as far as the understanding of matter is concerned, rendering the last three centuries of discussion a family quarrel. To be sure, we saw that Nagel is an exception from the rule in that he has paid some attention to panpsychism. But we found that Nagel fails to engage with the most systematically elaborated version of panspychism in modern thought, that of Alfred North Whitehead. The question is whether Nagel's reasons for dismissing panpsychism – or more precisely, for not being willing to endorse it – apply to Whitehead's defense of it.

Before we get to that, we shall look at Griffin's case for panpsychism, or what he prefers to call "panexperientialism", where the two basic features associated with mind – experience and spontaneity – are attributed to all units of nature. Griffin is quick to point out that, contrary to a common interpretation, panpsychism (the term he uses most of the time) is not committed to the view that, say, "rocks have feelings". The "pan" does not refer to all things existing but only to all genuine units or individuals. This is a very important point. Thus experience is not attributed to aggregational things, such as rocks and chairs, as such (Griffin 2007: 78). What Griffin does, then, is generalize *experience* to all individuals, whereas *consciousness*, by contrast, is understood as a very high level of experience. He quotes Whitehead's programmatic statement that "any doctrine which refuses to place human experience outside nature, must find in descriptions of human experience factors which also enter into the descriptions of less specialized natural occurrences. If there be no such factors, then the doctrine of human experience as a fact of nature is mere bluff". Therefore, we should "either admit dualism, at least as a provisional doctrine, or we should point out the identical elements connecting human experience with physical science" (Whitehead 1967 [1933]: 184f.; Griffin 2007: 79).

What is being challenged by panpsychism is nothing less than the fundamental ontological premise of the modern paradigm, a premise typically held onto even though the philosophical stalemate to which it has led brings its adherents to despair, as we noted above in the case of Colin McGinn – namely, the premise that the basic units of nature are material or physical in the sense of being devoid of all experience. Philosophers who, like Griffin, are today inclined to give the panpsychist alternative view the hearing so long denied it, typically share Bertrand Russell's sense, mentioned in Chapter 1, that – truth be told – we "know so little about the physical world". The intractability of the mind-body problem should accordingly be seen as following from the inadequacy of the dominant understanding of those phenomena that we are used to referring to as "physical".

The thought we encounter here is one that, surprisingly, was formulated by none less than Descartes himself, namely that there exists what might be called an epistemological asymmetry: we have a direct access to the mental that we lack in the case of the physical; hence, we are more likely to be wrong about the physical – about the nature of Descartes' "matter" – than about the experiential. Galen Strawson (2006), for example, holds that it is simply wrong to think that we have an adequate grasp of the fundamental nature of matter. Contra Searle and McGinn and the like, if there is a scandal in modern philosophy, it is not that we are still unable to explain the emergence of consciousness and mind, or to solve the mind-body problem. If anything, the scandal is the lack of recognition of the fact that *the failure to explain mind is a product of a failure to understand matter*.

Declaring the failure of the approaches loyal to the Cartesian notion of matter to be terminal, Griffin's recommendation is clear: always change a losing game. There is no use continuing to try and "beat the mind into the shape of the physical" (Nagel), understood in externalist categories. Panexperientialism as advocated by Griffin opens up a new strategy: "Begin with experience, which we *know* exists, and see if we can understand the various phenomena we call 'physical' in terms of varying degrees, organizations, and external perceptions of *it*" (2007: 91). Pursuing this strategy would have the benefit of avoiding a host of problems common to dualism and materialism. We would no longer have to think of experience as the "great exception", as amounting to an anomaly within the natural order, presumably because it is so utterly different from it. It would also free us from having to decide "at which arbitrary place to draw a line between experiencing and nonexperiencing individuals" (92).

We turn now to some of the main ideas elaborated in Whitehead's philosophy, which he himself came to designate as "process philosophy" and which Griffin regards as the most systematic and ambitious version of panpsychism ever developed.

Whitehead is not an easy read, a fact that may help explain the unwillingness of contemporary philosophers to engage with him, though it in no way justifies it. Indeed, the enduring reluctance among leading philosophers (including Nagel) to grant Whitehead's work the attention that it deserves is one of my reasons for discussing it so thoroughly in this chapter. To this day, Whitehead is probably most famous for having co-written, in 1910–13, *Principia Mathematica* with his erstwhile student, then Cambridge colleague Bertrand Russell. In 1925 he accepted a position at Harvard, and it was during this last phase of his career that he wrote the books most relevant to the discussion of panpsychism, most notably the complex and demanding *Process and Reality* (1929), subtitled – aptly for my purposes – *An Essay in Cosmology*.

Since what has led us to Whitehead is the suspicion that the notion of matter inaugurated by Descartes may be wrong, a good starting-point is his discussion of "the fallacy of misplaced concreteness" as presented in his Lowell Lectures at Harvard in 1925 and published as the monograph *Science and the Modern World*.

Looking back on the last three centuries of Western thought, Whitehead notes that "there persists . . . the fixed scientific cosmology which presupposes the ultimate fact of an irreducible brute matter, or material, spread throughout space in a flux of configurations". As taught be Descartes, "in itself such a material is senseless, valueless, purposeless" (Whitehead 1967 [1925]: 17). This is the assumption about matter that helped ground "scientific materialism" in the seventeenth century and that has ever since urged that we confine our inquiry to certain types of "facts", namely those epistemically accessed by way of "abstracting from the complete circumstances in which they occur" (ibid.). For a start, then, the fallacy Whitehead is concerned with has to do with the primacy of method over subject matter. In *Process and Reality*, he puts it like this: "The fallacy of misplaced concreteness consists in neglecting the degree of abstraction involved when an actual entity is considered merely so far as it exemplifies certain categories of thought. There are aspects of actualities which are simply ignored so long as we restrict thought to these categories" (1978 [1929]: 7f.).

To appreciate the significance of the fallacy, it is necessary to go beyond Griffin's presentation of it, based almost exclusively on Whitehead's 1925 lecture series. In *The Concept of Nature* (1920), Whitehead traces the origin of the fallacy as far back as to Aristotelian logic. He complains about "the muddle of importing the mere procedure of thought into the fact of nature", allowing the entity, "bared of all characteristics except those of space and time", to acquire "a physical status as the ultimate texture of nature; so that the course of nature is conceived as being merely the fortunes of matter in its adventure through space". In this way, philosophy "illegitimately transformed the bare entity, which is simply an abstraction necessary for the method of thought, into the metaphysical substratum of these factors in nature which in various senses are assigned to entities as their attributes". The muddle referred to prevents us from realizing that "it is impossible to express spatio-temporal truths without having recourse to relations involving relata other than bits of matter. [. . .] The true relata are events" (Whitehead 2007 [1920]: 20, 21, 24).

Another important instance of abstraction is at work in the theory of primary and secondary qualities going back to John Locke. Primary qualities are the attributes of matter which we do perceive and which are factually present in them, independently of our perceiving them, such as weight and size; or, in Descartes' version, figure, magnitude, and motion. Secondary qualities, by contrast, are other things which we perceive, such as colours, sounds, odors, tastes, and heat, which are *not* attributes of matter, "but are perceived by us as if they were such attributes" (2007: 27). Further qualities, such as purpose, volition, feelings – in short, the key elements of mentality, or psyche, or interiority – are also looked upon as perceived to be there, without actually being so; they are, in the wake of Descartes, dismissed as occult, as speculative and un- or even anti-scientific.

But why, asks Whitehead, do we perceive secondary qualities in the first place? His answer is not without sarcasm: "It seems an extremely unfortunate arrangement that we should perceive a lot of things that are not there". Whitehead regrets that there "now reigns in philosophy and in science an apathetic acquiescence in the conclusion that no coherent account can be given of nature as it is disclosed to us in sense-awareness, without dragging in its relations to mind" (ibid.). He gives no names, but I suspect he has in mind Kant as well as Descartes, Locke, and Hume. If so, his target is a broad one: empiricism, rationalism and idealism (in Kant's transcendental version). The consequences are serious: "The modern account of nature is not, as it should be, merely an account of what the mind knows of nature; but it is also confused with an account of what nature does to the mind. The result has been disastrous both to science and to philosophy, but chiefly to philosophy. It has transformed the grand question of the relations between nature and mind into the petty form of the interaction between body and mind" (ibid.). As Griffin misses no opportunity to observe, Whitehead was ahead of his time. Given the earlier discussion about why the mind-body problem has for so long dominated philosophy, to the detriment of more fundamental and far-reaching issues about the natural world, Whitehead's comment is as spot-on today as it was one hundred years ago.

Against this background, we can appreciate what Whitehead is addressing in his notion of the fallacy of misplaced concreteness. Much like in Hegel's logic and relational ontology, the fallacy is one of mistaking the abstract for the concrete, that is, assuming an abstraction from a concrete reality to be the totality, to capture what the thing at hand really is. We have been misled into assuming nature as it actually is to be composed of matter – bits of matter, at that – understood as having what Whitehead calls "simple location". Such location assumes the simple self-identity, the self-sufficiency in being what it is, of any here and any now, bared of anything that involves its spatio-temporal relations with anything else besides itself. As Whitehead points out, this way of thinking about things in nature, of identifying them and pronouncing on what they are and how they come to be so, would render induction impossible: "If in the location of configurations of matter throughout a stretch of time there is no inherent inference to any other times, past or future, it immediately follows that nature within any period of time does not refer to nature at any other period". In that case, "induction is not based on anything which can be observed as inherent in nature". Simple location, then, amounts to the denial – or removal by way of abstraction – of everything in the present fact which "inherently refers either to the past or to the future". As a result, "it looks as though memory, as well as induction, would fail to find any justification within nature itself" (1967: 51). In other words, simple location exemplifies what Whitehead refers to in his critique of Locke, namely the "extremely unfortunate arrangement that we should perceive a lot of things that are not there", the perception being debunked as invalid to the facts insofar as it is regarded as the product of projection: we "throw" qualities and properties at things which really are not there, failing to realize that what we regard as in-itself is instead things as they are for us, nature as it appears to mind, presumably, by way of being always already shaped by the workings of that mind, to allude to a figure of thought that is common to the two major epistemological rivals in modern philosophy, Hume and Kant. The upshot then is that "nature gets credit which should in truth be reserved for ourselves: the rose for its scent; the nightingale for his song; and the sun for his radiance". Whatever our experience to the contrary, the scientific truth is that "nature is a dull affair, soundless, scentless, colourless; merely the hurrying of material, endlessly, meaninglessly" (1967: 54).

In denying the concrete units of nature any inner duration, inner motion, and internal becoming, nature in its entirety is seen as made up of vacuous bits of matter with no internal values. This makes for a complete division between how we conceive – or indeed, experience – ourselves and how we regard nature as something alien to us, something we are set deeply ontologically apart from rather than being a part of. For example, we have memory, we relate to the past and anticipate the future; our capacities to situate our present situation in the temporality stretching out between a past and a future are constitutive for what the present, at any time, is for us, means to us. By contrast, natural units are denied any reference to the past as well as to the future. Whereas we take ourselves to be temporal beings through and through, and can only experience ourselves as such, the units of nature are said to be purely spatial: so many instantaneous bits of matter simply located in space (see Griffin 2007: 120f.). And again, the undeniable fact that we experience nature – spontaneously, affectively, through body and mind – as saturated with the "secondary qualities" that scientific naturalism is keen to remove from it, serves only as a stark reminder of the need to be suspicious of appearance as accessed by our senses – notoriously unreliable, the latter will only lead us astray, as none other than Descartes demonstrated in his *Meditations*, as far as scientific explanation of nature is concerned.

One might object that the critique Whitehead levels at the scientific philosophy that emerged in the seventeenth century is old hat, and that the premises upon which it is based have been abandoned, allowing for a much more complex understanding of nature. Whitehead disagrees: this system of concepts still reigns, still organizes scientific research; it even does so without any serious rival. And yet, "it is quite unbelievable" (1967: 24).

Matters are not that simple, though. According to one of his last works, *Modes of Thought*, published in 1938, thirteen years after the Lowell Lectures from which I have been quoting, developments in modern physics have undermined many of the key assumptions and axioms of the paradigm in question. It is now widely acknowledged that "there is no nature at an instant" (1968 [1938]: 146). The notion of nature as passive, of nature "at an instant", abstracted in the sense of simple location, has been replaced by the identification of matter with energy, "and energy is sheer activity" (137). Even so, the so-called new physics has not completely overcome the old dualism between experience and matter, regarded as mutually exclusive. Whitehead's observation in 1925 still holds: physics is stuck in abstraction. The science has no conceptual tools, no method, no ontology (be it implicit or explicit) that would enable it to recognize what anything "out there" is in itself, in its intrinsic reality. It continues to concentrate on entities only with regard to their extrinsic reality, considering doing so the hallmark of science, wholly in the spirit of Descartes (see 1967: 153).

III

So much for nature or "matter". What about mind?

According to Whitehead, misplaced concreteness renders not only our understanding of matter abstract; the same holds for mind. Since the fallacy concerns both, correcting Descartes' view on matter involves correcting his view on mind as well, even though Whitehead may be counted among the numerous philosophers today who judge that Descartes was first and foremost wrong about matter.

What is noteworthy about Whitehead's approach is his refusal to give separate analyses in this area: one devoted to mind and another devoted to matter. To get a sense of this, consider the following passage: "It is nonsense to ask if the colour red is real. The colour red is ingredient in the process of realization. The realities of nature are the prehensions in nature, that is to say, the events in nature" (1967: 72). Red is not something that can be accounted for mentalistically, something produced within a subject possessing a mind and projecting it (the colour red) onto an external world in itself devoid of colour. Indeed, as long as the mechanism of projection, taken as a deep-seated and largely unconscious disposition on the part of the human subject, is held onto as what explains the perception of colour (and similar qualities), then we will never get rid of the suspicion that the subject is projecting colour on what is in fact in itself a colourless object; that is, that the quality (property) in question really, and wholly, derives from the subject, not the object. Instead, Whitehead submits that red must be admitted as real, and as such, it must be fully situated in time and space. To get at this, Whitehead employs his technical term prehension, signifying "the essential unity of an event", that is to say, "the event as one entity, and not as a mere assemblage of parts or of ingredients". Space-time, he continues, is nothing else than "a system of pulling together of assemblages into unities" (ibid.). An event is temporal in all three dimensions:

it has a past, and it has a future; in its present, its here and now, it has "contemporaries" in that it "mirrors within itself the modes of its contemporaries as a display of immediate achievement" (ibid.).

Despite the anti-Cartesian thrust of his approach, Whitehead insists that it starts from the familiar Cartesian standpoint, "our own psychological field, as it stands for our cognition" (1967: 73). But whereas the Cartesian standpoint is suspicious to the point of blatant distrust (and existential despair!) as far as engaging with the "outer" reality is concerned, accessed by senses which may deceive us, thus privileging introspection as the road to epistemic certainty, Whitehead immediately situates cognition in a reality that is always ready to be shared by, partaken in, by mind and body alike. He promises to take our cognition for what it claims to be: "the self-knowledge of our bodily event", by which he means "the total event, and not the inspection of the details of the body" (ibid.). This self-knowledge is precisely not a narrow one, not a zooming-in on what would appear to be merely mine, mentalistically understood, as inhabiting an exclusively "inner" or "subjective" or even "private" world. On the contrary, self-knowledge as understood by Whitehead is eminently situated in something much wider than itself; it "discloses a prehensive unification of modal presences of entities beyond itself". What I do, then, says Whitehead, is "generalize by the use of the principle that this total bodily event is on the same level as all other events, except for an unusual complexity and stability of inherent pattern" (ibid.). In noting the "unusual complexity" that characterizes the human instantiation of the mind-body complex, Whitehead anticipates issues to be discussed later. For now, what matters is his bold conclusion: "[If] you start from the immediate facts of our psychological experience, as surely an empiricist should begin, you are at once led to the organic conception of nature of which the description has been commenced in this lecture" (ibid.).

The overall point, then, is that the scientific scheme handed down to us has failed on both counts, and for common reasons: *it has failed to explain mind no less than matter*. The scheme provides, Whitehead asserts, "none of those elements which compose the immediate psychological experiences of mankind" (1967: 73). Nor does it provide "any elementary trace of the organic unity of a whole, from which the organic unities of electrons, protons, molecules, and living bodies can emerge". The principal explanatory poverty of that scheme is that it acknowledges "no reason in the nature of things why portions of material should have any physical relations to each other" (ibid.). Whitehead makes the same point with regard to events: "the relatednesses of an event are all internal relations" (1967: 123). Each relationship enters into the essence of the event; "apart from that relationship, the event would not be itself". In stressing this point, Whitehead flatly contradicts the view dominant since Descartes and Hobbes, that spatio-temporal relationships are external (ibid.). What Descartes described as primary attributes of physical bodies, are in fact the forms of internal relationships between actual occasions, and within them. As we will examine in more detail below, what is at issue here is nothing less than the shift from materialism to organism as the basic idea of physical science – the shift Whitehead was to advance most systematically in *Process and Reality*, seeking to displace a system of thought by basing nature upon the (Cartesian) concept of matter with one basing with a system that bases nature on organism.

Before we assess the implications of that shift of paradigm, however, we need to take a closer look at Whitehead's understanding of mind.

The principle Whitehead is propagating is that "consciousness presupposes experiences, and not experience consciousness". This means that "an actual entity may, or may not, be conscious of some part of its experience. Its experience is its complete, formal constitution, including its consciousness, if any" (1978 [1929]: 53). Opposing the theory of representative perception advanced by Descartes and Locke, Whitehead construes perception to mean "taking account of the essential character of the thing perceived", whereby what is truly important is that "we certainly do take account of things of which at the same time we have no explicit cognition". Indeed, common usage as well as philosophical doctrines are both guilty of attributing excessive cognition to perception and apprehension alike. In Whitehead's terminology, prehension designates "uncognitive apprehension" (1967: 69). The upshot is that the unity of a moment of experience is not a product of, nor is it dependent on, conscious operations – the latter come later, as it were, giving the lie to our tendency, psychologically no less than epistemologically, to put them at the start of the sequence.

Whitehead sets out to repudiate the assumption, implicit in the philosophical tradition, that the basic elements of experience are to be described in terms of one, or all, of the three ingredients: consciousness, thought, and sense-perception. The last term is used in the sense familiar to us from Hume and Locke: "conscious perception in the mode of presentational immediacy". In practice, observes Whitehead, sense-perception is narrowed down to visual perception, taken as the mode of experiencing the world, of having access to whatever is given "out there", that is the point of departure for knowledge. It thus enjoys primacy over all other modes. But according to the philosophy of organism, this is simply wrong. The three mentioned components are in fact "unessential elements in experience, either physical or mental". This is because "any instance of experience is dipolar, whether that instance be God or an actual occasion of the world" (1978: 36).

We will return to the point about dipolarity (or bipolarity). In accepting William James' view that consciousness is a particular *function* of experience, Whitehead warns against seeing consciousness as "a preexistent stuff lying in waiting, as it were, to be filled by this content or that", to quote Griffin's depiction (2007: 132). We must likewise reject the common assumption that "those elements that are most clearly lit up by consciousness must be the elements that actually arise first in experience" (ibid.). The opposite is closer to the truth, according to Whitehead, since "consciousness only arises in a late derivate phase of complex integration", it tends to illuminate the data of that late phase, not the initial data (1978: 162). Failing to recognize this has produced "most of the difficulties of philosophy", contends Whitehead. That experience has been explained in a topsy-turvy fashion; the wrong end first is brought out in the case of Hume in particular, where emotional and purposeful experience have been held to follow from impressions of sensation. The point is that "the order of dawning, clearly and distinctly, in consciousness is not the order of metaphysical priority" (ibid.).

This insight may explain Whitehead's reasons for holding that the still-dominant view of "mind" and "matter" as stark opposites exemplifies the fallacy of mistaking the abstract for the concrete. Hume is a case in point, holding as primitive what is in fact derivative in the sequence of experience. His overall thesis, on which his entire epistemology is founded, is that sensory perception is *the* primary element in our experience. Whitehead considers this to be wrong: our perceptual experience does *not* begin with sensory perception. Rather – and here I rely on Griffin's exposition – sensory perception is derivate from complex to simpler modes of perception, the first of which is "perception in the mode of causal efficacy". It is through this that "we know most of those things that we inevitably presuppose in practice", what Griffin refers to as "hard-core common-sense notions", such as the reality of the external world (which Hume confessed he could not help assuming in practice, although doing so contradicted his philosophical skepticism); the reality of the past and the future and, therefore, of time; the reality of our conscious experience, and the reality of bodily influence on conscious experience and the efficacy of conscious experience for bodily behavior (see Griffin 2007: 34ff.). For my present purposes, however, the most important one of the common-sense notions that are hard-core in that we in practice cannot function without them, has to do with the reality of *efficient causation* understood as the real influence of one thing (or many things) on another.

Contradicting the widespread view that Hume and Kant were in conflict over causation, Whitehead faults both for making the same mistake, namely inverting the true nature of experience: for both, "presentational immediacy was the primary fact of perception, and any apprehension of causality was, somehow or other, to be elicited from this primary fact" (1978: 173). Their view of the relation between causation and perception, understood as items in the epistemic subject's experience, can be traced to Locke and Descartes, although it is Kant in particular who made it into what Whitehead calls "the Neolithic weapon of 'critical' philosophy". The underlying fallacy is the one we discussed above: taking clearness in consciousness as "evidence for primitiveness in the genetic process" (ibid.).

As is well known, Hume explains our assumption that causality is at work in the outside world, as a presumably necessary connection between the behavior of (say, billiard ball) A and the behavior "caused" in B, as a psychological effect of the repetition of associated presentational experiences: every time we see that A, we expect that B will follow. Observed regularity in repetition translates into an assumption about a necessary connection between the phenomena involved in the sequence, taking A as the cause and B as the effect. In other words, a psychological process in the observing subject is made into a subject-independent truth about the relationship between the phenomena observed. Hence, "the subjective response is the beginning and end of all that there is to be said about causation"; "the response is response to presentational immediacy, and to nothing else" (1978: 174).

Whitehead offers an example to show why Hume is mistaken: In the dark, the electric light is suddenly turned on, and the man's eyes blink. Whitehead's detailed analysis of the blinking man's experience is highly instructive and merits quoting in full:

> The sequence of percepts, in the mode of presentational immediacy, is flash of light, feeling of eye-closure, instant of darkness. The three are practically simultaneous; though the flash maintains its priority over the other two, and these two latter precepts are indistinguishable as to priority. According to the philosophy of organism, the man also experiences another percept in the mode of causal efficacy. He feels that the experiences of the *eye* in the matter of the flash are causal of the blink. The man himself will have no doubt of it.

> In fact, it is the feeling of causality which enables the man to distinguish the priority of the flash; and the inversion of the argument, whereby the temporal sequence "flash to blink" is made the premise for the "causality" belief, has its origin in pure theory. The man will explain the experience by saying, "The flash made me blink"; and if this statement be doubted, he will reply, "I know it, because I felt it".
>
> (1978: 175)

Whitehead holds that we should accept the man's statement, that the flash *made* him blink, and that this demonstrates causal efficacy. But Hume cannot accept it. Hume would insist that "what the man really felt was his *habit* of blinking after flashes". But how, asks Whitehead, can a "habit" be felt, when a "cause" cannot be felt? Is there any presentational immediacy in the feeling of a habit that is invoked by Hume? By a sleight of hand, Whitehead argues, Hume "confuses a 'habit of feeling blinks after flashes' with a *feeling of the habit* of feeling blinks after flashes" (1978: 175).

As Alastair Hannay (2018) has pointed out, it was Hume's theory of association that gave him the idea of there being a (necessary, causal) connection between phenomena or events; what we develop as subjects is the habit of assuming such connection to be operative in the outside world, effectively at work there, between the phenomena themselves, without any part played by us. By contrast, Whitehead's position is that causality, far from being a "subjective" notion – though one shared universally among subjects – without foothold in the goings-on in the world, pervades everything. If we want to know how the notion of causality arouse, the answer is that it did so – necessarily so – "because mankind lives amid experiences in the mode of causal efficacy" (1978: 175). While this claim is no doubt in keeping with Whitehead's position, it must be said that neither the example he gives about the blinking man nor his very general contention about causality "pervading everything" suffice to provide an alternative answer to the ones given by Hume and Kant respectively regarding the validity of the notion of causal *necessity* – i.e., that there is a necessary, as opposed to contingent, connection between phenomena perceived as distinct in the world, one deemed as cause and the other as its unavoidable and invariable consequence.

On Whitehead's analysis, perception in the mode of causal efficacy is a nonsensory mode of perception that is more basic than sensory perception. This is so is often overlooked in philosophy, yet confirmed in our daily life where we perceive our body and mind as a unity, not as divided: "Our feeling of bodily unity is a primary experience. It is an experience so habitual and so completely a matter of course that we rarely mention it. No one ever says, Here am I, and I have brought my body with me" (1968 [1938]: 114). Our sense of unity with the body has "the same original as our sense of unity with our immediate past of personal experience. It is another case of nonsensuous perception" (1966: 189).

The failure to recognize causal efficacy as operative on the basic level of nonsensuous perception is one Whitehead traces to the errors he sees Locke and Hume as committing in their sensationalist doctrine. The first error is the assumption of a few "definite avenues of communication with the external world, the five sense-organs". This leads to the further assumption that the search for the data is to be narrowed to the question, "what data are directly provided by the activity of the sense-organs", with pride of place being given to the eyes (1966 [1933]: 225).

Whitehead grants that this doctrine has a general truth to it. It is evidently important for our practical affairs as well as in scientific observation. Nevertheless, it fails to grasp our primary mode of orienting ourselves in the world. Whitehead's way of putting this is to point out that "the living organ of experience is the living body as a whole". Hence every instability on the part of the body – be it chemical, physical, or molar – "imposes an activity of readjustment throughout the whole organism" (1966: 225).

The second error is one associated with Descartes in particular: the assumption that "the sole way of examining experience is by acts of conscious introspective analysis". The attitude of introspection has in fact led us astray instead of securing the access to indubitable truth about how we experience and gain knowledge about the world that Descartes promised in giving it methodological primacy. According to Whitehead, introspection "lifts the clear-cut data of sensation into primacy, and cloaks the vague compulsions and derivations which form the main stuff of experience. In particular it rules out that immediate sense of derivation from the body, which is the reason for our instinctive identification of our bodies with ourselves" (1966: 226).

We can now appreciate what Whitehead is up to when, in concluding his critical discussion of Locke in *Process and Reality*, he states that "the philosophy of organism abolishes the detached mind" (1978: 56). Doing so is all about giving *mental activity* a deeper as well as more expansive meaning than what has been common in Western thought: such activity, and the capacities required to exhibit it, is not the prerogative of human subjects but must be recognized as a property of everything that qualifies as real in Whitehead's sense: *mental activity, that is, is operative in what Descartes categorized as matter as well as in mind* (meaning the mind of humans only in Descartes' case). Elaborating the declaration to abolish the detached mind, wanting to situate it everywhere, Whitehead writes that "mental activity is one of the modes of feeling belonging to all actual events in some degree, but only amounting to conscious intellectuality in some actual entities. This higher grade of mental activity is the intellectual self-analysis of the entity in an earlier stage of incompletion, effected by intellectual feelings produced in a later stage of concrescence" (1978: 56).

I should point out that "mind" is not Whitehead's chosen term for making this bold claim. How far "away" from humans, and how far "down" the evolutionary ladder he is willing to go when it comes to acknowledging that mind – or mind-like capacities – is at work, is hard to tell. But it is beyond dispute that his philosophy of organism, especially as developed and defended against major philosophical rivals such as Descartes, Locke, Hume, and Kant in *Process and Reality*, aims for a radical break with the dominant epistemologies. As Griffin's exposition makes clear, Whitehead to a large extent rests his case on his attempt to show how wrong his predecessors have been in privileging presentational immediacy in (human) experience, and in equating it with sensory perception. This significance of this mistake for the radical break Whitehead aims for is brought out in the assumption that it has led to: that entities without sensory organs can have no perceptual experience at all. Griffin observes that even if most philosophers will allow some form of experience to most animals, they draw the line at the point where there seems to be no sensory organs. However, Griffin notes, "if presentational immediacy and therefore sensory perception are derivative forms of perception even in us, then it is not impossible in principle to generalize some kind of perceptual experience to all individuals, *however primitive*" (2007: 136; italics mine). To back up this reading, Griffin quotes the following from Whitehead: "We must assign the mode of causal efficacy to the fundamental constitution of an occasion so that in germ this mode belongs even to organisms of the lowest grade; while the mode of presentational immediacy requires the more sophistical activity of the later stages of process, so as to belong only to organisms of a relatively high grade" (1978: 172).

In dismissing as false the Humean notion that causation depends on vivid sense data, Whitehead urges that we recognize causal efficacy as at work in our experience as consisting of emotional, appetitive, and purposive responses to the world around us as well as to our body and our own past. On Griffin's reading, conceiving experience in this way helps provide "elements some faint analogy to which can less implausibly be ascribed all the way down" (2007: 136). I must admit wanting a stronger argument than this rather circumspect and indirect one to justify the ascription of such experience "all the way down", hence expanding the loci of experience far beyond (or beneath) humans.

At this point Griffin recalls Whitehead's programmatic statement in *Adventures of Ideas* (quoted earlier) that "Any doctrine which refuses to place human experience outside nature, must find in descriptions of human experience factors which also enter into the description of less specialized natural occurrences. If there be no such factors, the doctrine of human experience as a fact within nature is mere bluff", leading Whitehead to conclude that "we should either admit dualism, or we should point out the identical elements connecting human experience with physical science" (1966: 185). Laying out these identical elements in great detail is exactly what Whitehead sets out to do, especially in the dense phenomenological sections of *Process and Reality*.

It is important to keep in mind that on Whitehead's analysis, the denial of experience in (nonhuman) nature has a very specific origin: the twin fallacies of simple location and of misplaced concreteness mislead us into portraying matter as having none but spatial properties; "spatializing nature" is what presentational immediacy brings about, and our allegedly deep-seated "disposition" to access and conceive of nature by way of this immediacy has for centuries been both justified and culturally entrenched by prominent philosophers. As we have seen, granting privilege to presentational immediacy in sensory perception twice over, as it were – in terms of generic primitiveness as well as in terms of justification (validity) – has resulted in a near-total neglect of the perceptual mode of causal effect, testifying as it does to a very different view of nature, one giving the lie to the whole post-Cartesian undertaking of denuding – emptying – nature of mentality, interiority, feeling, and purposiveness – qualities amounting to the "experience" that we have seen Whitehead hold to be operative in nonhuman nature, more primitive and basic than the consciousness found in humans.

Let me sum up. Whitehead agrees with the modern paradigm of scientific materialism that the fundamental units of nature are material (physical). But he rejects the notion that in being physical, they are devoid of all experience. To the contrary, the units of nature are shot through with experience: they are what they are and are sustained as such in virtue of being instances of experience, carriers and transmitters of experience among themselves, relationally as well as internally. In short, experience goes all the way down.

What does this mean?

IV

In *Science and the Modern World*, Whitehead declares the physicists' notion of energy to be an abstraction. Following the hypothesis proposed by quantum theory that the ultimate elements of matter are in their essence vibratory, the field, says Whitehead, is open for the introduction of some new doctrine of organism to displace the Cartesian view of matter as devoid of experience. But why does he describe the notion of energy as "obviously an abstraction"? His immediate answer is that "the concrete fact, which is the organism, must be a complete expression of the character of a real occurrence" (1967: 36). But what, more precisely, is the energy described in physics an abstraction from?

A more comprehensive answer is offered in the important chapter "Objects and subjects" in *Adventures of Ideas*. "The notion of physical energy, which is at the base of physics, must be conceived as an abstraction from the complex energy, emotional and purposeful, inherent in the subjective form of the final synthesis in which each occasion completes itself" (1966: 154). For all the progress promised by its explicit rejection of the Cartesian notion of matter, then, the widespread idea in modern physics that energy is the ultimate reality embodied in all actual things is another instance of the fallacy of misplaced concreteness. True, as Griffin (2007: 154) comments, this concept does point to something real; even so, it is an abstraction from full-fledged creative experience as understood by Whitehead, which is always emotional and purposeful, a matter of spontaneity, self-creation, and self-enjoyment. It is an often-made complaint in Whitehead's writings that physics – regardless of the distinction between old and new – abstracts from "what anything is in itself", taken as its intrinsic reality. Owing to the abstractions practiced by physicists old and new, and without which they perhaps would not be able to "do physics" as the science they take it to be, we are (still) not successful in grasping the full reality of the most elementary types of actual events at the base of nature. To make some headway here, aspiring to do justice to what the actual events are in themselves, Griffin elaborates what he calls an "inner physics" meant to be wholly in the spirit of Whitehead's philosophy of organism.

In setting out to "enlarge" the notion of energy, Griffin stresses that this is no mere conceptual issue: it is substantial in that (so the thesis goes) there are two phases to the embodiment of energy in any event: the subjective phase and the objective phase. This entails that "psychic energy" ceases to have a purely metaphorical meaning; it is not – no longer, in keeping with traditional physics – confined to the kind of psychic energy known in our own experience in its "subjective" or intrinsic phase. The development of an inner biology of cells – referred to as "psychological physiology" by Whitehead – will involve "positing a subjective as well as an objective phase of the embodiment of energy in *all* unified events", so that "the transition from intrinsic or psychic energy to *extrinsic* energy will be assumed to be going on all the time" (2007: 154). Griffin holds Whitehead to have articulated the same thought when stating that "the energetic activity considered in physics is the emotional intensity entertained in life" (1968: 168). Enlarging the notion of energy, then, means to expand its scope all across the board and all the way down; it means to argue its presence in all entities insofar as they qualify as real, as actual events in Whitehead's terminology. Griffin sees this bold thesis of his supported in the books Whitehead authored in his late years, spent at Harvard and often referred to as his "metaphysical" phase.

We take physics to study the simplest things in nature, the elements that cannot be divided into yet smaller parts and from which the complexity of the physical world as a whole has emerged. As Griffin notes, the so-called elementary particles, such as photons, protons, neutrons, and neutrinos, or more recently quarks, as described by physics, are assumed to be the simplest actual things. He then points out that Whitehead disagreed, asserting that the simplest *actual* things are the simplest occasions of experience, of which the elementary particles as described by physicists are abstractions. Importantly, on the view advanced by Whitehead, "these simplest things" are the most "widespread habits of nature", by which he means what have commonly been called "laws of nature" (1968: 154f.). In employing the term "habit", Whitehead betrays his debt to the pragmatist philosophy of William James. The shift from thinking about laws of nature as a matter of regularities considered to result from supernatural imposition, to thinking of them as habits is a truly far-reaching one according to Griffin.

One of the poverties of the sensationalist doctrine advanced by Locke and Hume is that there is no way to conceptualize what our actual experience seems to suggest, namely that "efficient causation conditions final causation, which then becomes the basis for another act of efficient causation". Griffin illustrates what he takes Whitehead to have in mind by offering the following example:

> My present experience is conditioned by causation both from my body and from past states of my own mind. These efficient causes, however, do not totally determine my present experience: I still can, and in fact must, decide precisely how to respond to those conditioning causes – those bodily cravings, those promises made, those plans, those sensory precepts. When I do make my decisions, they seem to exert causal efficacy on my bodily states and my own subsequent experiences, and so on. The idea that our *stream* of experience is really composed of momentary *occasions* of experience, each of which begins with physical experience and ends with a mental reaction hereto, explains how efficient and final causation can be interwoven.
>
> (Griffin 2007: 158)

This is a fine example as far as it goes. But in my view it does not go far enough. Everything Griffin says here about the interplay, the dynamic back-and-forth between efficient and final causation, seems to me to be fully in keeping with Whitehead's position, just as Griffin claims it to be. The problem is that the interplay and the way in which it plays out is described *in one particular case*, that of a human subject, easily recognizable as such by all of us. But what about all the other cases – the other-than-human ones – for which Whitehead also, in his bold thesis (and this of course is very much to do with its appearing so bold), claims that the same sort of interplay between efficient and final cause takes place? Otherwise put, if Griffin wants to make good on his Whitehead-inspired thesis to enlarge the notion of psychic energy – enlarge it substantially, as a matter of its factual scope and extension, as opposed to just conceptually – would he not have to give examples taken from the domain of reality (of Whiteheadean actual events) besides the human one? Indeed, would not the radicalism of Whitehead's philosophy of organism demand that one is willing to *abandon* the very privileging – by way of example, illustration, instantiation – of the human case that Griffin is reproducing from the very school of thought he sets out to repudiate?

V

Griffin's chief objective is to elaborate an "inner physics" based on Whitehead. The term is not Whitehead's, however. In *Process and Reality*, Whitehead uses "physical physiology" to deal with the subservient inorganic apparatus, and "psychological" to deal with "entirely living" nexus, "partly in abstraction from the inorganic apparatus, and partly in respect to their response to the inorganic apparatus, and partly in regard to their response to each other" (1978: 103). He notes that "a structured society may have more or less 'life', and that there is no absolute gap between 'living' and 'non-living' societies" (102). Whitehead elaborates in terms familiar from the science of ecology: "A complex inorganic system of interaction is built up for the protection of the 'entirely living' nexus, and the originative actions of the living elements are protective of the whole system". At the same time, "the reactions of the whole system provide the intimate environment required by the 'entirely living' nexus". Importantly, "we do not know of any living society devoid of its subservient apparatus of inorganic societies" (103).

Although Whitehead did not himself use the term "inner physics", he did suggest the need to complement the "physical physiology" practiced thus far with a "psychological physiology", noting that the latter is still very much in the process of incubation (1978: 103). To explain what is at issue here, Griffin invokes Nagel's concept of "subjective universals" (discussed in Chapter 1; see Nagel 1986: 21), referring to general features of subjective experience and raising the question: what sort of entities exhibit subjective experience? The subjective universals are meant to apply to all subjects, understood as momentary occasions of experience, from the human level to the actualities studied by physics. But who are these subjects? What does it require to qualify as one? Griffin points out that not all identifiable entities in the world qualify; many of these, such as "rocks, lakes, and computers, have a merely aggregational, not a subjective or experiential, unity". Subjective universals, then, apply only to all "genuine individuals", whether simple or compound individuals. *Which* things are to be considered true individuals is an empirical question, Griffin tells us. What is decisive is whether the behavior of the entity in question suggests "a unity of responsive action that involves an element of spontaneity", meaning that the response does not seem "fully explainable in terms of efficient causation from prior events" (2007: 151). Griffin observes that Whitehead himself supposed humans, most other animals, eukaryotic cells, molecules, and atoms to be compound individuals, whereas subatomic (elementary) particles are regarded as primary individuals.

Subjective universals explicate what is implicit in the idea that "all individuals embody creative experience", which as such is not an actuality but is that which is exemplified in all actualities, as brought out in Whitehead's phrase that creative experience is the "universal of universals characterizing ultimate matters of fact" (1978: 21). Creativity, Whitehead emphasizes, is the principle of novelty. In the highest experiences, the "subjective forms" (which according to Griffin equal Nagel's subjective universals) may include consciousness, as exhibited in humans. "Emotional forms", however, are included in experiences at all levels, recalling the basic role. Whitehead accords to feeling in the sense of physical prehension: All experience – throughout the entire continuum from the lowest to the highest level – begins with feelings or prehensions of other actual things, in which they grasp aspects of those things. Hence, the experience of the efficacy of other things (for good or bad) for itself is what commences an experience in each particular case. Besides his strong emphasis of the foundation part playing by feeling, Whitehead – as we saw above – stresses that of final causation or self-determination: the experience is partly *causa sui*; it does not create itself out of nothing, *ex nihilo*, but out of the physical experiences imposed on it by its past. The element of self-determination, called "subjective aim" by Whitehead, varies from being trivial or slight in the most elementary experiences to being pervasive in the higher subjective forms, as expressed in conscious purposes. The aim is executed in the "exclusion of the boundless wealth of alternative potentiality" and in the "definite factor of novelty" (1968: 152).

Even though Griffin does a good job in laying out what a Whiteheadean "inner physics" amounts to, his understanding of "subjective universals" raises a critical question. Remember his list of the sorts of entities that do not qualify: rocks, lakes, and computers, all of them having "a merely aggregational, not a subjective or experiential, unity" (Griffin 2007: 151). If a major reason for engaging with Whitehead's philosophy of organism is that it provides a much richer and more differentiated ontology than do classical idealism and materialism, as well as present-day physicalism, then it seems strange to run together such different entities as rocks, lakes, and computers, portraying all three as essentially alike both in terms of what they lack and in terms of what they have. A rock is in inanimate physical thing; a lake is a complex organic ecosystem; a computer is a human-made artifact. What brings them into existence in the first place, what sustains them over time, and what decides the functions they fulfill and how well and for how long they will do so, greatly varies between the three, as does the role of various types of *other* entities in helping shape, channel, and determine that whole trajectory as an actual entity among – and co-constituted by – myriads of others in the world. The three entities listed by Griffin to exemplify similarity in fact attest to the opposite, difference, which spans the whole range from a site of life (the lake) to a man-made machine (the computer), that is, from what is eminently organic to what is wholly artificial, from what exists in virtue of growing to what exists in virtue of having been made (fabricated), hence from what has an irreducibly internal origin to what has an external one.

It is easy to understand Griffin's wish to draw the line somewhere: only on pain of implausibility would we postulate *all* existing entities as subjects in the sense intended by Nagel's subjective universals. But the question is whether such reluctance to go all the way is one Griffin is entitled to if he is to live up to Whitehead's dictum that either we must admit dualism, or we must affirm "the identical elements connecting human experience with physical science" (1966: 184). There is no denying it: Whitehead's thesis is truly radical in holding that "each actuality [not some of them] is essentially bipolar, physical and mental", whereby "the integration of the physical and mental side into a unity of experience is a self-formation which is a process of concrescence" (1978: 108). He hastens to add that "though mentality is non-spatial, mentality is always a reaction from, and integration with, physical experience which is spatial. It is obvious that we must not demand another mentality presiding over these other actualities (a kind of Uncle Sam, over above all the U.S. citizens)" (ibid.).

I admit that my argument falls short of repudiating the point that Griffin is making: that the three entities lack a subjective or experiential unity, having a merely aggregational one. But on the Whiteheadean premises, Griffin emphatically affirms that it appears true that a lake – just like a rock and a computer – lacks in capacities (or in parts exhibiting capacities) manifesting a subjective or experiential unity. Whether unity of *that kind* can be attributed to a lake, as (analytically) opposed to its various parts (i.e., a multitude of interdependent marine species), is difficult, or downright impossible, to answer, at least if asked in the Nagelian fashion: "What is it like to be a lake?" Even so, there is a strong sense in which that very question seems much *closer* to having an affirmative answer in the case of the lake than in that of the rock or the computer, again underscoring the ontological difference between them that Griffin ignores. So while it may be tempting to agree with Griffin's statement that "which things are to be considered true individuals is an empirical question", it does not sit well with Whiteheads' uncompromising metaphysical zeal.

Central to Whitehead's "inner physics" is his critique of "enduring individuals", traditionally conceived as numerically self-identical substances enduring through time. He reconceives such individuals as "temporally ordered *societies* of momentary occasions of experience, in which there is a *perpetual oscillation* between subjectivity and objectivity, final causation and efficient causation" (Griffin 2007: 159). I consider Whitehead's own way of putting this as unmistakably ecological: Each occasion's "activity in *self-formation* passes into its activity of *other*-formation"; it is "transcended by the creative urge that as such belongs to the essential constitution of each such occasion". The occasion, then, "arises as an effect facing its past and ends as a cause facing its future". In between, Whitehead states, "there lies the teleology of the Universe" (1966: 193, 194).

Griffin's endeavor to portray Whitehead's position as an "inner physics" involves two generalizations: first, that of momentary events from physics to the psyche, and second, that of the perpetual oscillation between subjectivity and objectivity to all enduring individuals, including those of physics. This yields the following account:

> Each event is a subject for itself before it is an object for others. All things other than our own experience *appear* to be mere objects, rather than subjects, because by the time they can be prehended they *are* objects; their subjectivity has perished. This is the very precondition for their being objects for our perception, or for their exerting any efficient causation whatsoever. This is one of the several interrelated explanations as to why, if the universe is *really* composed exhaustively of active subjects, it *seems* to be composed primarily of passive objects. It is *not* composed exhaustively of things that are *simply* subjects (as was the Leibnizian universe); it is composed of subjects-that-become-objects. So, we are right to think that everything that we perceive is an object – in the ontological as well as the epistemic sense of the term. We are only wrong to think of them as *mere* objects.
>
> (Griffin 2007: 159)

Several things are worth noting here. Unlike Whitehead, Griffin throughout employs the subject-object distinction to make his case, associating – wholly conventionally – subject with agency, mentality, and purposefulness, object with the assumed opposite properties of passivity and exteriority. He then uses the distinction to explain why it is that "we" (human subjects) tend to perceive as objects those that are also subjects: in the very act of perception of objects other-than-us, we deny them their inner agency and everything that goes with it, reserving it for ourselves only.

If we grant, for the sake of the argument, that this is the truth about what we do, why do we do it? For psychological reasons, in the sense of a deep-seated, universal disposition in all human subjects, along the lines suggested by Hume or Nietzsche? Or for cultural reasons, having to do with the ambition of the society to which we belong to deny other-than-human creatures agency so as to be able to exploit them with impunity? But is the most stubborn notion of our tendency not to commit a fallacy with respect to where to draw the subject/object distinction the very opposite one, that of anthropomorphizing in the sense of attributing agency to beings (objects) factually devoid of it, falsely perceiving objects as subjects rather than the other way around?

Griffin does not raise these questions. Doing so would have clarified the basis on which he argues for an "inner physics". It would also help remove the ambiguity raised by his cited formulation that "All things other than our own experience *appear* to be mere objects, rather than subjects, because by the time they can be prehended they *are* objects; their subjectivity has perished" (2007: 159). Has it really? Is this Whitehead's position? Or has it perished for us only, not in – and for – the things themselves? Moreover, is Griffin's use of the term "object" such that in having *anything* except our own experience as an object of our experience, we automatically and per definition objectify it in the sense of denying it subjectivity understood as capacities for agency? If so, what about how we experience the experience of other human subjects?

In recalling Whitehead's remark that "one task of a sound metaphysics is to exhibit final and efficient causes in their proper relation to each other" (1978: 84; Griffin 2007: 157), Griffin gains a good opportunity to address these questions. Whitehead's view, as we know, is that there is an interplay between final and efficient causes in every entity, the embarrassment produced by the legacy of Cartesianism being that the one type of cause is associated with mind (or consciousness) and the other with matter (or body), resulting in a failure to account for their interaction. This being so, Griffin is correct in pointing out that "there was no way to conceptualize what our experience seems to suggest, that efficient causation conditions final causation, which then becomes the basis for another act of efficient causation" (2007: 158). Importantly, the efficient causes, in the form of causation both from my body and from past states of my own mind, do not totally determine my present experience: "I still can, and in fact must, decide precisely how to respond to those conditioning causes – those bodily cravings, those promises made, those plans, those sensory percepts" (ibid.).

There is something remarkably conventional about Griffin's approach here, contrasting sharply with the radicalness of the thesis that his declared "Whiteheadean panexperientialism" commits him to: that "experiential features belong to all actual entities as fully as do those objective features that are usually called 'physical'" (2007: 160). This thesis has a vital implication as far as final and efficient causes are concerned: namely, that both are present in all actual entities, and that in being so they relate, and they interact. The originality and boldness of the Whiteheadean thesis consist in holding such interaction between the two types of causes to occur in the vast richness of nonhuman entities in addition to in humans – *within and throughout* the vast ontological region that Descartes called matter just as in what he called mind, reserving the latter for humans only.

My point is a simple one: Griffin cannot make good on the thesis to which he subscribes as long as he sticks to orchestrating his discussion and selecting his examples solely within the framework of human experiences. He must transcend that framework, as did Whitehead. It does not help that the human case is the one that originally, and notoriously, undermined the Cartesian position, falsifying it by purely internal means and for purely internal reasons, as it were. The failure to explain the inter-relationship and mutual interplay between what was posited as absolutely separate and so mutually exclusive – mind and body (matter), final and efficient causes – may well have been, and has hitherto been widely seen as, *particularly* embarrassing in that it so unavoidably pertains to the human case. It is the scandal of not being able to account for the interplay in what is considered as its prime case, the one for which that account is absolutely imperative. Whitehead disagrees: the nonhuman instantiations of the interplay between the two types of causes are no less important, no less philosophically crucial, no less in demand of a way of thinking about the interplay, explaining it and doing justice to it than is the human case so long, and so wrongfully, treated as paradigmatic.

What is praiseworthy about Griffin at this point is that he clearly wants to pursue his proposed "inner physics" at an *ontological* level, as distinct from an epistemological one – although, as we have seen, the (Humean) presuppositions dominating the latter prove difficult to escape.

To appreciate what is at issue, Griffin argues that Whitehead's philosophy shows the way in giving us "theoretical homogeneity for all individuals, from photons to cells to human beings". What allows for, even calls for, such homogeneity is nothing merely conceptual: the sense in which photons and human beings are ontologically similar does so. In other words, the question about whether we may speak all across the board, all the way down, is answered in the affirmative: we may. The thesis of homogeneity opposes, on the one hand, the idealist view that physical predicates are less real than experiential ones, and on the other hand, the materialist one that experiential predicates are less real than the physical. Affirming homogeneity entails that "*every actual entity has two modes of existence*, a subjective mode, in which it has none but experiential properties, and an objective or superjective mode, in which it has none but objective properties" in a sense that is not limited to properties observable through sensory perception (to recall Whitehead's critique of Hume) (Griffin 2007: 161).

The big question is: does the homogeneity thesis imply panpsychism?

VI

Griffin turns to Galen Strawson's suggestion that "being experience-involving is a fundamental property of existing things on a par with extension, rest mass, or electric charge" (quoted in 2007: 160). Recall that the Whiteheadean "panexperientialism" that Griffin defends "does say that experiential features belong to all actual entities as fully as do those objective features that are usually called 'physical'. In this sense they are 'on a par' with them" (2007: 160). Nevertheless, Griffin warns against looking upon experiential phenomena as "just one more variety" of physical phenomena. The point is that they do not belong to the actualities "in the same mode of existence", that is to say, on a par with properties such as mass, charge, and spatial extension. Hoping to make the idea of their equal reality intelligible, Griffin states that "the experiential predicates apply only to the actual entity in its subjective mode of existence, when it exists in and for itself, whereas the other predicates apply to it only in its objective or superjective mode of existence, when it exists for others (as an efficient cause on them)" (ibid.). However, Griffin finds problematic, even "perhaps impossible to conceive consistently", the Spinozist notion that "the experiential and the objective features of an actual entity exist *simultaneously*". Why? Because it would "make the relation between final causation (self-determination) and efficient causation unintelligible: How could an actual entity already be exerting efficient causation on others while it was still determining exactly what it is to be" (ibid.). Dismissing Spinozist panpsychists on the grounds that they are "generally determinists as they cannot attribute any degree of self-determining power to any individuals", Griffin sums up his "Whiteheadean" view to the effect that "the objective mode of an actual entity, with its objective properties, exists only after the subjective mode has come to completion" (ibid.).

I cannot go into the differences that no doubt exist between Spinoza's doctrine of "psycho-physical parallelism" and Whitehead's philosophy of organism. The issue of freedom raised by Griffin as crucial to these differences is a huge topic in its own right. Followers of Spinoza would be quick to point out that his type of panpsychism is not committed to rejecting freedom but to show how it is inseparable from necessity, the two being not opposites, mutually exclusive at that, but being two sides of the same coin, the same reality, when correctly understood. Leaving Spinoza aside, the more specific problem that I see with Griffin's argument concerns the precise ontological status of his talk about "determining what it is to be". Who, or what, is he holding to do – to be able to – such determining? The question is a simple one, and should meet with a simple answer. But it does not. Why?

The problem encountered here goes back to the generalizations we saw Griffin take Whitehead to be promoting, laying the foundation for a so-called "inner physics". The main idea is that, to quote Griffin's gloss on what he clearly takes to be Whitehead's view, "the type of causality we experience in relation to our own past experiences and our bodily members can also be generalized to other things" (2007: 160). This claim prompts the simple question: can it?

Consider how Griffin makes the claim: "The way in which our present experience prehends immediately previous occasions of our own experience, incorporating their basic character and continuing their projects and subjective forms, can be used to understand the continuity of enduring individuals in the worlds of biology and physics" (2007: 160f.). To be sure, the phrase "can be used to understand" does not sound very contentious. Far from articulating a bold claim, it seems to express a truism: how, except by using our understanding of our own experience, would we be able to form a view about what pertains to "enduring individuals in the world of biology and physics"?

The tough question is whether this truism warrants the larger claim Griffin is making – the far from commonsensical panexperientialist one that "experiential features belong to all actual entities". As I pointed out in the Introduction, it is a (hermeneutical) fact that in trying to understand how experiential properties are operative in other-than-human entities, we draw upon our own such properties qua humans. But our non-optional doing so does not in itself justify the *results* of our doing so: it does not demonstrate the correctness and validity of what we take ourselves to understand about enduring individuals in the worlds of biology and physics. To address what holds for such individuals in terms of *their* experiential features – as distinct from ours, from the ones we draw on in trying to understand – they themselves would have to constitute the material for inquiry.

Griffin supports his claim by stating that "my present experience is not *simply* a continuation of my past experiences, with their emotions and purposes, but is constantly broken into by multiple routes of causation from my bodily members, some of which carry causal influence from things beyond my body". Expanding the origin and scope of causal influence in this way, as it applies to the human subject, Griffin goes on to suggest that "this fact can be generalized to understand many-termed causal relations *in nature in general*, the relations that generate space as well as time" (2007: 161; my italics). He uses this point for all it is worth: following, "the principle that our own experience is part of nature as much as anything else, we can generalize from our own distinctively mental experience, in which we grasp possibilities as such with appetition, to the notion that all unitary events have a mental dimension to their experience, hence at least some degree of final causation" (ibid.).

But Griffin moves too fast. He draws substantial conclusions about the worlds of biology and physics, and specifically about the experiential features of the sorts of entities that comprise them, from what he must admit is our "distinctive" mental experience as humans. Again, how can we determine whether the conclusions drawn from an admittedly human-experiential viewpoint are *true to the facts about the other-than-human beings they concern*?

Griffin backs up the generalization that he is advancing by saying that it amounts to an inference justified in Whitehead's "genuine nondualism". He cites Whitehead's asking the following questions: "But why not construe the later forms [of evolution] by analogy to the earlier forms? Why not reverse the process? It would seem to be more sensible, more truly empirical, to allow each living species to make its own contribution to the demonstration of factors inherent in living things" (Whitehead 1929: 11, quoted in Griffin 2007: 161).

To be sure, plenty of passages in Whitehead seem to support Griffin's use of generalization. For example, in the chapter "Organisms and Environment" in *Process and Reality*, Whitehead observes that "in describing the capacities, realized or unrealized, of an actual occasion, we have, with Locke, taken human experience as an example upon which to found the generalized description required for metaphysics". However, he hastens to add that "when we turn to the lower organisms we have first to determine which among such capacities fade from realization into irrelevance, that is to say, by comparison with human experience which is our standard" (1978: 112).

Again, one may comment, "how could it be otherwise?" But the simplicity of the comment should not deceive us. *What* precisely is it that could not have been otherwise? We need to make a distinction conspicuously absent in Griffin's discussion: between human experience as the point of view required for understanding – making claims about – anything at all, in Gadamer's hermeneutical sense (1960), and human experience as the model with reference to which experience in other-than-human entities are acknowledged and assessed – if at all.

Whereas Griffin tends to have only the hermeneutical dimension in mind, to the detriment of the substantial one, Whitehead explicitly addresses the latter when he notes – critically – that "in any metaphysical scheme founded upon Kantian or Hegelian traditions, experience is the product of operations which lie among the higher of the human modes of functioning. For such schemes, ordered experience is the result of schematization of modes of *thought*, concerning causation, substance, quality, quantity" (1978: 113). In short, experiential unity is conceived as attained in the guise of modes of thought.

By contrast, the philosophy of organism sets out to construct "a critique of pure feeling". In doing so, Whitehead descends from the higher organisms to the lower; the more we do so, the more we have to abandon the human-specific model as paradigmatic for experience, keeping in mind that, even in the human case, the long-standing equation of experience with "clarity of knowledge" is mistaken. In *The Function of Reason*, influenced by William James' notion of "thick experience", Whitehead stresses that "in our own lives, and at any moment, there is a focus of attention, a few items in clarity of awareness, but interconnected vaguely and yet insistently with other items in dim apprehension, and this dimness shading off imperceptibly into undiscriminated feeling" (1929: 63). At a primitive level, experience is not a prerogative of thought, of beings demonstrably capable of thought; it is a matter of feeling. Whitehead's way of making this point is cryptic, however. The critique of pure feeling, he says, "should supersede the remaining *Critiques* required in the Kantian philosophy". The implication is that "in the organic philosophy Kant's 'Transcendental Aesthetic' becomes a distorted fragment of what should have been his main topic" (1978: 113).

Return now to Griffin and what he views as the crux of Whitehead's position: that Whitehead's distinct manner of overcoming the vicious dualism between minds as temporal but nonspatial and physical things as spatial but not essentially temporal consists in "*putting a duality within each actual entity*" (2007: 161).

To be sure, this is a crucial idea in *Process and Reality*. To explain how Whitehead arrived at it, Griffin quotes what may well be one of the most important passages in the book:

> Each actuality is essentially bipolar, physical and mental, and the physical inheritance is essentially accompanied by a conceptual reaction partly conformed to it, and partly introductory of a relevant novel contrast, but always introducing emphasis, valuation and purpose. The integration of the physical and mental side into a unity of experience is a self-formation which is a process of concrescence and which by the principle of objective immortality characterizes the creativity which transcends it. So though mentality is nonspatial, mentality is always a reaction from, and integration with, physical experience which is spatial.
>
> (Whitehead 1978: 108)

Given the self-conscious radicalness of Whitehead's philosophy, it may come as a surprise to learn that he agrees that the physical is spatial and the mental nonspatial, which on the face of it is the conventional view. Nonetheless, Griffin maintains, the originality of Whitehead's contribution has to do with his success in avoiding "a vicious dualism between two different types of actual things" (2007: 162). How, more precisely, does Whitehead succeed here? Griffin answers by referring to Whitehead's doctrine of momentary occasions of experience, "each of which is both subject and object, and thereby both with and without duration, and each of which as subject is both physical and mental and thereby both spatial and nonspatial". This doctrine, Griffin tells us, is Whitehead's way of overcoming the fallacy of misplaced concreteness and its "resulting vicious dualism between mind and matter, soul and body" (ibid.).

Stating what Whitehead set out to achieve is one thing. Showing that he is successful is another. To be sure, there is much to admire about Griffin's pedagogical skills in explaining what Whitehead is trying to accomplish in *Process and Reality*, "a book that in richness and suggestiveness is unsurpassed, but in opacity is monumental" (to quote Donald Sherburne [1981: 2]). Now consider this explanatory formulation on Griffin's part: "We know ourselves from within, hence as having duration, and other things from without, hence as devoid of duration. To translate this epistemic duality into an ontological dualism between two different kinds of actualities – those that are always subjects and those that are always objects – is to commit a category mistake" (2007: 161f.). The mistake is to "contrast things as known from within with things as known from without and to conclude from this *epistemic* contrast that they are *ontologically* disparate" (ibid.).

This is probably as straightforward a rendition of one of Whitehead's most crucial claims as one can hope for. The problem, though, is the same as in my above discussion: a bold thesis about how "each actual entity" is – namely, a duality in the given sense – is exemplified as well as backed up by an appeal to how "we" know ourselves from within and from without, only thereupon to – falsely, the thesis has it – translate this epistemic duality (i.e., the duality we are committed to authoring as epistemic subjects) into an ontological dualism (i.e., a division between two substantially different kinds of actualities in the real world). To point out, as does Griffin, on Whitehead's behalf, that we as epistemic subjects have a deep-seated disposition to commit this category mistake, is to make an important point about *us* (one also made, in different versions, by Hume and Kant). But it is not to make a compelling point about *the world of actual entities, including the manifold of other-than-human such*, entities for every one of which an ontological internal duality is claimed on the thesis that Griffin aspires to defend.

My point is that what the thesis *requires* is an account on the level of the lowest forms of mental experience, so as to show, *as exhibited there*, the bipolarity of the mental and the physical that is meant in holding that "experiential features belong to all actual entities", thus including the lowest forms as well as the highest in terms of (degree, sophistication of) mentality. But what is being *offered* so as to argue the thesis is something else. In Griffin's case, it is the observation, reiterated again and again, that "we never come upon an unexperienced thing; we never catch things as they are apart from being experienced" (2007: 175). That may well be so, in the sense that, hermeneutically, we never "come upon" anything at all in the world without it being us who do so, namely, by way of making each thing that we encounter into an object of our experience (however dim and lacking in sustained focus and clarity it may be in the individual case, to allude to Whitehead's warning against equating experience with knowledge).

True, then, we never come upon an unexperienced thing. In the sense just given, this is trivially, yet importantly, true. Hume and Kant would both concur. But is there not an ambiguity to the notion "unexperienced thing"? Is it not the deeper question that Griffin is required to do justice to the ontological one, as analytically distinct from the epistemic one – ontological in the sense that what we really need to investigate to make good on the thesis is the thing not as experienced (by us) but *the thing as itself experiencing*, as exhibiting experience and hence in its capacity as subject not object?

It is highly regrettable that Griffin fails to raise this question. What he does offer is a distinction between what he calls a good and a bad type of "subjectivism". Subjectivism in the bad sense is solipsism, denying – or being unable to account for – the experience of other actual entities as actual. Subjectivism in the good sense is "the recognition that our own subjective experience is what is most immediately given to us and as such should be the starting point of our analysis of what actual things are really like" (2007: 173). On the basis of what Griffin depicts as "a thoroughgoing empiricist starting point", "we have no basis for taking as actualities anything other than the experiences themselves" (ibid.). What does he mean? He answers by stating that "we have no direct experiential basis for saying that there are any vacuous actualities, actualities devoid of experience" (ibid.). Maybe not. But this is not a strong argument. What needs to be argued for, to justify the panexperiential thesis, is the opposite: *the existence of actualities – besides humans, and comprising physical reality in its entirety – that exhibit experience in the sense of inhabiting a mental pole*. Griffin's follow-up to the formulation, which cited that we have "no empirical basis for assigning a meaning to the assertion that there are actual things devoid of experience, because we can give no hint as to what they might be *in themselves*" (ibid.), is an argument that cuts both ways: if the fact (presumably not to be altered) is that we cannot access what the actual things that we are talking about are "in themselves", then the case that states that they possess experience (i.e., are loci of experience) seems just as groundless or purely speculative as the case that states that they do not.

Griffin's own conclusion from the cited inalterable fact is that "all other actualities affirmed must be conceived by analogy with a moment of our own experience". To this effect, he cites Whitehead's statement that "the percipient occasion is its own standard of actuality" (1978: 145; Griffin 2007: 176). Griffin then says that "if our own experience is our very standard of actuality, then every actuality is to be conceived as a unification of a given multiplicity into an experiential unity", taking this to be supported by Whitehead's view that experience means the "self-enjoyment of being one among many, and of being one arising out of the composition of many" (ibid.).

There is no denying it: the claims made are of enormous generality and presented on a breathtakingly high level of abstraction. What precisely are we referring to in talking about "every actuality"? What exemplifies "the one" and what would exemplify "the many"? Griffin's pointing out that our own experience "obviously should be taken as an example of an actuality" is not helpful. That is not what is at issue and what his thesis requires him to show. Examples of actualities *other* than ourselves, *yet* exhibiting experience (in some minimal form and degree as demanded by the thesis of internal duality as found in "all" actualities) are required.

To his credit, Griffin concedes that the argument that he is making is circular. But he is quick to present this, if not exactly as a virtue, than at least as a matter of necessity. "Any ontology", he assures us, "is built on a similar circularity", by which he means that "something is taken as the paradigm case of actuality, from which it follows that that something is an instance of actuality" (2007: 176). This point only takes us so far. Griffin's argument, made on behalf of Whitehead's position, suffers from excessive generality at the very point where precise content and concrete examples are called for.

There is also an expositional oddity to be mentioned. We saw that Griffin quoted Whitehead's statement that "the percipient occasion is its own standard of actuality" (1978: 145), taking this to imply that our own experience is the standard for actuality as such. Consulting the passage from which the quote is taken, however, shows Whitehead to be speaking not about our experience about but the philosophy of organism: "For the philosophy of organism, the percipient occasion is its own standard of actuality. If in its knowledge other actual entities appear, it can only be because they conform to its standard of actuality. There can only be evidence of a world of actual entities, if the immediate actual entity discloses them as essential to its own composition" (1978: 145).

VII

Let me summarize, in a way independent of the specifics of Griffin's exposition, what I consider to be the core of Whitehead's notion of "inner physics".

There is a duality in all actual entities. Each and every one of them inhabit a mental pole and a physical pole, or a subjective pole and an objective pole, if you will. The objective pole testifies to efficient causation, the subjective pole to final causation. This does not means that the two properties – the mental and the physical – at all times and in all entities show up in equal measure, or with the same degree of complexity. That is not the case. What is the case is that there is the potential for both poles being exhibited, but the mental is more complex and sophisticated in some entities than in others: more so in humans, less so in animals, still less in plants, and so on down to the "lowest", most primitive level of physical reality. For all the significance and variance of this differentiation from the highest to the lowest, however, both poles go to the very bottom, to the micro physical level, and their trajectory – which is one of interaction between them – has to be traced all along to way. Whitehead is not particularly interested in atoms, however. But he does say that there is no way that we can determine what kind of experientiality or mental life is exhibited at the lowest level of entities. His preferred procedure is to proceed from the macro to the micro, not the other way around. What we can say, most generally, is that what begins with physical experience (objective pole, efficient cause) – ends with mental experience (subjective pole, final cause), of which human consciousness – owing to its degree of spontaneity, of creativity, of self-determination, of subjective aim, in short: of everything we associate with freedom – is the highest form. But again, this should not lead us to conclude that in the entities at the lowest level there is "only" the objective pole, and in those at the highest that there is "only" the subjective pole. On the contrary: the duality, the two-in-one, is to be found throughout – horizontally and vertically, if you will – in everything that goes to make up physical reality, persistently at that, thereby crossing the mind/matter divide erected by Cartesian dualism, as well as rejecting the two variants of monism that have tried – unsuccessfully – to overcome it, namely idealism in the form of the primacy of mind and materialism in the form of the primacy of matter.

Griffin has done an important job in explaining how Whitehead turned the traditional dualism between physicality and mentality into a duality, in the form of bipolarity, *within* each actual occasion. His demonstration of how we tend to conclude, from the epistemic contrast between things as known from within and things as known from without, that they are ontologically disparate, is also of great value for assessing the importance of Whitehead's position. That said, there is a sense that Griffin makes too much of this particular idea in Whitehead's philosophy of organism. After all, it figures far less prominently in *Process and Reality* than Griffin's readers are led to think. And while the passages where Whitehead does address the inner duality (bipolarity) of the physical and the mental are indeed bold and suggestive, they are lacking in clarity: so much more is being claimed than is being shown.

Several commentators have argued that Whitehead made his clearest statement in a lecture entitled "Time", given in 1926, and containing the often-cited idea that "the mental occasion supercedes the physical occasion". On David Skrbina's reading, this shows that Whitehead regarded every natural occurrence as "a complete event in which an initial physical phase (pole) is replaced by a mental phase" (Skrbina 2007: 176). At a minimum, then, Skrbina assures us, Whitehead considered "all realities as events" and held that "all events have both physical and mental aspects to them". In maintaining that "every actuality has mentality", Whitehead is seen as affirming panpsychism (ibid.).

I believe that Griffin and Skrbina are justified in viewing Whitehead as a proponent of panpsychism. While it may be uncontroversial to hold this to be true in a general way, the issue of panpsychism as articulated in late Whitehead becomes complicated once we start looking in his work for a clear answer to the question of what precisely his panpsychism amounts to. We have seen this in my discussion of Griffin's interpretation of Whitehead. And we witness it again, in a somewhat different manner, in the case of Skrbina, who, like Griffin, approaches Whitehead from a professed pro-panpsychist standpoint.

In Skrbina's case, the point can be put fairly simply: if he is correct in saying that according to Whitehead "every natural occurrence is a complete event in which an initial physical phase (pole) is replaced by a mental phase", it is questionable whether this – as Skrbina claims – affirms the panpsychist thesis that every actuality has mentality. The answer hinges on what is meant by holding that an initial physical phase is being *replaced* by a mental phase. Is there first only, and wholly, a physical phase (pole), subsequently to be transformed into a mental one, eventually what is only and wholly a mental phase (pole), so that at the beginning there is physicality only and at the end of the process there is mentality only? Would putting it like this hold true for *some* entities, but not for others, thus marking a limit to attempts to generalize all across the board (as panpsychists are inclined to do)? If true, what about the point made above in the discussion of Griffin, stressing the importance of *interplay*, of interaction and inter-relatedness, between the two poles, postulated as pervading each actuality – simply in virtue of being real – throughout its entire trajectory, from beginning to end? In the simplest possible terms: Is not the gist of the panpsychist thesis that there is – always, everywhere – a question of both-and (both physicality and mentality) as far as all actualities are concerned, and never of either the one or the other? Is not the signature thesis of panpsychism precisely that co-existence – co-habitation – of physicality and mentality is the main characteristic of reality as such?

The problem is not that there will be cases where physicality predominates over mentality, and others where mentality predominates over physicality, as illustrated by the continuum from the amoeba to the human (to stick to organic life). We grant that mentality, as far as its degree, richness, and sophistication are concerned, is much more predominant in the case of the human than the amoeba. The hard question has to do with the exact status of the mental pole in the lowest regions of the hierarchy: is there mentality there, from the very beginning, or is there not? Sure, *knowing* that there is, and being able to prove it beyond dispute, so that we all are convinced, may well be impossible, simply too tall an order (more on this below). As for now, the task is to reach a conclusion about how to understand Whitehead's position.

Skrbina considers Whitehead to be "somewhat clearer" in his late book *Modes of Thought* (1938) than in *Process and Reality*. To illustrate, he cites the following statement from the chapter entitled "Nature Alive": "[T]his sharp division between mentality and nature has no ground in our fundamental observation. [. . .] I conclude that we should conceive mental operations as among the factors which make up the constitution of nature" (1968: 156). As it stands, however, and as Skrbina lets it stand, this statement is much too sweeping to establish the clarity we are looking for.

A passage a few pages later, not cited by Skrbina, throws more light on the matter. Having noted that physical science has "reduced nature to activity", Whitehead comments that "the fundamental question remains: How do we add content to the notion of bare activity?", suggesting that this question can only be answered by "fusing life with nature". What does he mean? He says that, in the first place, we must "distinguish life from mentality". The sort of functioning he terms "conceptual experience" is "the entertainment of possibilities for ideal realization in abstraction from any sheer physical realization. The most obvious example of conceptual experience is the entertainment of alternatives. Life lies below this grade of mentality. Life is the enjoyment of emotion, derived from the past and aimed at the future" (1968: 166f.). Importantly, in stating that "life lies beyond this grade of mentality", Whitehead does not suggest that there is life *devoid* of mentality, but that there is life that does not exhibit conceptual experience understood as the ability (however rudimentary) to entertain alternatives. Does Whitehead mean to imply here that below a certain grade of mentality as inhabited in life, there is no subjective aim and no final causation? If so, does this entail a concession to critics of panpsychism? And how does what Whitehead say here sit with his claim in *Process and Reality* (quoted earlier) that "mental activity is one of the modes of feeling belonging to all actual entities in some degree, but only amounting to conscious intellectuality in some actual entities" (1978: 56)? Or with the even more sweeping and bold claim that "to be actual must mean that all actual things are alike objects, enjoying objective immortality in fashioning creative actions; and that all actual things are subjects, each prehending the universe from which it arises" (ibid.).

VIII

In *The Phenomenon of Life*, Hans Jonas raises a central issue not addressed by Griffin or Skrbina. Although crediting Whitehead's project with a philosophical importance "unequaled in our time", Jonas has serious doubts about whether it represents a gain. Writes Jonas: "The blurring of the difference between inanimate and animate nature, which the spread of inwardness down to the physical foundation entails, seems a high price to pay for the claimed atomicity of 'actual occasions' which requires this universal inwardness to obviate their otherwise inevitable monistic isolation" (Jonas 2001: 96). The result, Jonas continues, is a "submersion of discontinuity where it matters – between life and nonlife – against its injection where it is hypothetical – between phases of physical duration" (ibid.). Furthermore, the polarity of being and not-being according to Jonas is not accounted for in Whitehead's system, and therefore is not the phenomenon of death. Thus, "the deep anxiety of biological existence has no place in the magnificent scheme". What we get from Whitehead, then, is a story of "intrinsically secured success: all becoming is self-realization, each event is in itself complete (or it would not be actual), each perishing is a seal on the fact of completion achieved" (ibid.). The sting of death has been removed from the order of nature.

Whitehead's claim that "there is no absolute gap between 'living' and 'nonliving' societies" (1978: 102), and that "we do not know of any living society devoid of its subservient apparatus of inorganic societies" (103), may well be affirmed by more biologists and philosophers today than when he made it nearly a hundred years ago, in 1929. And as we will see in a moment, Whitehead recognizes that there is such a thing as a "risky environment", one upon which all entities – at least all living such – rely. Nonetheless, Jonas has a valid point:

Whitehead concentrates on what is required for any given entity to persist in its existence, regarding its success in doing so to be in each case an achievement – yet he largely takes it for granted that persistence will be secured, as shown in his notion of endurance. The precariousness of the entire undertaking of persistence seems largely lost on Whitehead, be it on the level of individuals or of wholes (to be sure, the two interact). Today, we may say that death, not to mention extinction, understood as what undercuts persistence, plays a bigger role than it did not only for Whitehead but for Jonas as well, writing in 1966. When Whitehead observed that "any physical object which by its influence deteriorates its environment, commits suicide" (1967: 109), he did so in the calm and lofty theoretical context of elucidating the mechanism of evolution. Little could he know that, a couple of generations later, with the advent of the Anthropocene, the object on path to committing such suicide would be humankind, dramatically reinforcing Jonas' point about the reality of death and indicating that the primacy of life, the achievement of persistence, can no longer be taken for granted. We shall address the significance of this in Chapters 4 and 5.

IX

Before bringing my discussion to an end, there is one further topic that we need to explore, that of value.

Value is not among the topics that Whitehead himself selected for systematic investigation in *Process and Reality*. Even so, the "inner physics" Griffin takes that work to propagate cannot be fully understood without highlighting Whitehead's original view of value.

Whitehead's most elaborate discussion of value occurs in his *Science and the Modern World*. The context is Wordsworth's worship of nature as brought out in his poems. The nature-poetry of the romantic, Whitehead writes, was "a protest on behalf of the organic view of nature"; specifically, it was "a protest against the exclusion of value from the essence of matter of fact" (1967: 94).

But there is more to it than the case made for value-in-nature in the romantic protest against the meaninglessness and alienation produced by the post-Cartesian removal of value from nature. There is the larger case to be made against the received wisdom that value is a prerogative of humans, and doubly so: value originates in humans and in humans only in the sense of our supposedly unique capacity for attributing it to selected items, such attribution being seen as warranted only when its object is another human person, making for a wholly anthropocentric outlook concerning genesis as well as extension.

As much as Whitehead shares the complaint voiced by Wordsworth and the romantic movement, his reasons for doing so are to a large extent an integral part of his philosophy of organism. To appreciate this, consider Whitehead's understanding of what it means for something to endure. As Isabelle Stengers observes in her monumental *Thinking with Whitehead*, if an order of nature as conceived by Whitehead exists, it is because the event, enacting as it does a unifying grasp in the way explained in *Process and Reality*, can also "hold fast and endure through an environment that never ceases to change". Endurance, that is to say, is not only a fact: it is an accomplishment and an achievement. It requires, Stengers submits, that "the order of nature integrate what we are accustomed to thinking as paradigmatically human, that is, value" (2011: 156).

At this point, Stengers quotes almost *in extenso* the seminal paragraph in *Science and the Modern World* from which Griffin, revealing his surprisingly modest interest in the topic, cites all-too restrictedly and only in passing:

> Remembering the poetic rendering of our concrete experience, we see at once that the element of value, of being valuable, of having value, of being an end in itself, of being something which is for its own sake, must not be omitted in any account of an event as the most concrete actual something. "Value" is the word I use for the intrinsic reality of an event. Value is an element which permeates through and through the poetic view of nature. We have only to transfer to the very texture of realization in that value which we recognize so readily in terms of human life. This is the secret of Wordsworth's worship of nature. Realisation therefore is in itself an attainment of value. But there is no such thing as mere value. Value is the outcome of limitation. The definite finite entity is the selected mode which is the shaping of achievement; apart from such shaping into individual matter of fact there is no attainment. The mere fusion of all that there is would be the nonentity of indefiniteness. The salvation of reality is in its obstinate, irreducible, matter-of-fact-entities, which are limited to being no other than themselves. Neither science, nor art, nor creative action can tear itself away from obstinate, irreducible, limited facts. The endurance of things has its significance in the self-retention of that which imposes itself as a definite attainment for its own sake. That which endures is limited, obstructive, intolerant, infecting its environment with its

> own aspects. But it is not self-sufficient. The aspects of all things enter into its very nature. It is only itself as drawing together into its own limitation the larger whole in which it finds itself. Conversely it is only itself by lending its aspects to this same environment in which it finds itself. The problem of evolution is the development of enduring harmonies of enduring shapes of value, which merge into higher attainments of things beyond themselves.
>
> (Whitehead 1967: 93f.)

Considering the importance of the claims about value made in this extraordinarily rich passage, it is surprising that Whitehead did not use his most systematic work, *Process and Reality*, to further elaborate them and show their full implications for his philosophy of organism.

In holding value to be "the intrinsic reality of an event" and that "realization is in itself the attainment of value", Whitehead signals a clear-cut breach with what to this day remains the most common understanding of value, in philosophy as well as in common sense: no human agent, no value. Value, that is to say, has no independent ontological reality or standing: short of being *in re*, in the things themselves as a crucial component of what they are, of their intrinsic mode of existence. Value is held to be wholly and exhaustedly *for-us* in that it is something we attribute to (a selected class of) things: we are producers and projectors of value, and for lack of our being so, there would be no value to talk about. Remove value-conferring human agents from the world, and you remover value's sole source as well as exclusive locus in the same stroke.

Freeing value from its anthropocentric prison and thus from any psychological connotation, Whitehead submits that if we, as humans, are different from everything else, it is not because we nourish values. Stengers' gloss on Whitehead's position is that "value belongs to the order of nature; it is what is realized by all that exists, in the sense that what exists succeeds in enduring, succeeds in maintaining its individual way of gathering together, that is, of making things hold together in a determinate way. Value indicates a success in and of itself" (2011: 157).

There is an unmistakable holistic, ecological ring to what Whitehead says about value, though his commentators, Stengers no less than Griffin, for some reason do not use the term and omit to pursue its relevance, Whitehead's for ecological thinking as well as ecology's for Whitehead's philosophy of organism. It is a sadly missed opportunity.

In saying that "realization is in itself the attainment of value", and that value indicates a success in and for itself, Whitehead's understanding of value recalls the notion of "conatus" propagated by Spinoza, highlighting that everything that exists is committed to a striving to persist in that existence. "Conatus", however, is not the term employed by Whitehead. His chosen terms are endurance and infection. Endurance, as Stengers emphasizes, is not an attribute of the individual who endures, but depends first of all on a dynamics of infection: "All that succeeds in enduring has succeeded in infecting its environment in a way that is compatible with this endurance" (2011: 157). "Infection" in Whitehead captures what the poets, led by Wordsworth, celebrate as "presence". Celebration is indeed the appropriate term here since it is a poet's experience that, owing to his or her openness toward everything encountered in nature, is infected by, say, the mountain, allowing it to shape the perceptions, feelings, thought, and imagination that being in its presence entails. More generally, infection as understood by Whitehead designates the success constituted by all endurances in a changing world, whereby that success relies on and is inseparable from an "infectious holding-together that is not a fusion but a valorization, a determinate shaping, conferring a value – that is, a role – on what is prehended". Far from being a mere quantity, the value of a variable "presupposes the stability of the role that one thing plays for something else and measures the importance of that role". As Stengers proceeds to remark, infection as intended in Whitehead's sense of "attributing a role to something" is anything but neutral: it can be all-decisive, as brought out in her example: "The role you attribute to me drives me crazy" (2011: 158).

The dynamics of infection finds privileged examples in *relations of dependence*, underscoring Whitehead's view that the idea of autonomy, taken to mean perfect self-sufficiency in existing, is obsolete. Stengers writes that living beings, as studied in biology, have "the privilege of exhibiting endurance *qua* achievement, in a risky environment". Biologists, that is, "address the enduring organism insofar as its stability is at stake for this organism itself, insofar as it *matters*" (2011: 173).

This statement requires explanation. To have something – anything – matter, must presuppose some elemental capacity for exhibiting value, in that, again in a very elementary sense, the mattering – the making a difference – will amount to its being either conducive or threatening to the endurance, understood as persistence, of the entity in question, endurance being an achievement "in the midst of a risky environment", as we saw Stengers observe. For something to matter, then, means for it to be exempt from neutrality, that is, to be either positive or negative for the entity and its future trajectory. To be sure, this is again a clearly ecological way of making the point. Whitehead, however, goes beyond that in *Process and Reality* where value is "no longer associated with the organism insofar as it endures, but with its actualization as such", following Stengers' attempt to reconstruct what remains "enigmatic" in Whitehead, who, as she laconically remarks, "acts as though what he writes is clear" (2011: 207, 206, 205).

The crucial term for what Whitehead is up to here is "envisagement", signifying, among other things, "a process without a subject"; envisagement, explains Stengers, "translates what must be attributed to the underlying activity itself if actuality is to be value. And this concerns every actuality" (2011: 208). This is a truly radical thesis, one for which we are not totally unprepared, of course: it means that, in the intended metaphysical sense, envisagement "confers no privilege upon human envisaging experience" (ibid.). Thus, Whitehead submits that "every occasion, even the most tenuous, fugitive wrinkle differentiating general energy, has a value *qua* individualization of the eternal activity, that is, it constitutes an individual purpose". It follows that "all the aspects of our most singular, most routine experiences, must be able to be described as exhibiting that which, from the metaphysical standpoint, is presupposed by all actualization" (ibid.). What Whitehead aims to do, then, is to identify value and actuality. And if only actuality has value, Stengers explains, "it is value as our experience can manifest it, with the sense of novelty, of risk, of purpose, all of which allude to what Whitehead calls 'addition to the future', not a value that would reduce our experience to a fiction or an illusion" (2011: 220).

Stengers' 500-page exposition of Whitehead's work – concentrating on his so-called "metaphysical" phase that commenced when he, at the age of 64, accepted the professorship offered him at Harvard in 1925 – is relevant for my purposes in that she dwells on the one topic in Whitehead's metaphysics that Griffin so conspicuously neglects: value. Notwithstanding the superiority of her approach when it comes to assessing the role played by value in Whitehead's late philosophy, Stengers tends to do as did Griffin before her: having duly highlighted the true radicalness and speculative thrust of Whitehead's main thesis, she proceeds to explain and exemplify it by invoking a number of experiences that "we" have, attesting to the validity of Whitehead's often highly idiosyncratic concepts in the familiar domain demarcated by all-too human experiences, immediately recognizable as such, thus in effect lending plausibility to what Whitehead is suggesting.

Yet Whitehead is suggesting so much more: that is the whole point of his metaphysical undertaking. The novel terms he invents, and the familiar ones he gives a novel meaning, such as "infection", "prehension", "eternal objects", "envisagement", and – last but not least – "value", are not predominantly meant to apply to the domain of human experiences but to the vast and multifaceted total reality within which those experiences are situated: the extra-human one. That is the gist of his thesis, cited by Stengers, that value is intrinsic to *every* occasion, every actual entity. Stengers moves abruptly from Whitehead's general claim that the dynamics of infection finds privileged examples in relations of dependency to allusions about "what makes human beings hold together"; from his claim that an entity "can only succeed in holding fast if the way it infects its environment does not give rise to repercussions that make it lose its hold" to asking the question "how can I keep hold if the person I am dealing with provides no foothold, does not integrate within her own reality the aspect of herself I am proposing her"; from his claim that infection marks that there is nothing neutral about "attributing a role to something" to the observation that "the role you attribute to me drives me crazy" (2011: 158, 159).

To be sure, in all these cases we immediately grasp what Stengers is saying: the examples she provides from our shared repertoire of human, all-too-human experiences – about dependence, about being shaped and determined by others, human others – are striking and spot-on. But the plausibility that they gain and the stamp of approval that we are willing to give them are forthcoming on false grounds and at a high price: that of deviating from the incomparably vaster domain – i.e., ontological terrain – with regard to which Whitehead set out to claim the validity of the concepts in question. Insofar as Whitehead's metaphysical thesis confers no privilege upon human experience, the case for the thesis cannot be made by invoking a series of examples taken from such experience.

3 Prospects and pitfalls of agential realism

Karen Barad's *Meeting the Universe Halfway: Quantum Physics and the Entanglement of Matter and Meaning* is one of the most thought-provoking philosophy books published in the last decade. As the title suggests, it covers a lot of territory in a number of fields, exploring fundamental issues in the philosophy of science, in epistemology, and in ontology, profiting throughout from Barad's background in theoretical particle physics. Although her book does not discuss Whitehead, we shall see that Barad engages with some of the same questions as did he, approaching them not through a philosophy of organism but via the famous debate between the giants of the "new physics": Albert Einstein, Werner Heisenberg and Niels Bohr. In showing the relevance of Bohr's contribution in particular, which frames her proposal for an "agential realism", Barad brings the revolutionary insights of quantum physics to bear on the issues of the "nature of nature", of matter, and of mind. In doing so, she makes her own contribution to the discourse of posthumanism that will occupy us for the rest of the present work.

I

Central to Barad's project is the discussion between Heisenberg and Bohr about the "uncertainty principle". Coined by Heisenberg, the principle concerns the limits to our knowledge of the behavior of physical objects, like atoms or electrons; more specifically, there is a necessary limit to what we can simultaneously know about certain pairs of physical quantities, such as the position and momentum of a particle. The point, then, is not that we cannot have any knowledge about a particle's position and momentum, but that there is a trade-off concerning "how well we can know both quantities *at once*: the more we know about a particle's position, the less we know about its momentum, and vice versa" (Barad 2007: 7).

Bohr differed from Heisenberg in holding that the deeper issue involved here is not that of a limit to what we can *know* about the position and momentum of a particle simultaneously; rather, it is that particles do not *have* determinate values of position and momentum simultaneously. Heisenberg had worried that in measuring any of the characteristics of a particle, we cannot avoid disturbing its premeasurement values; therefore, the more we know about a particle's position, the less we will know about its momentum, and vice versa. As Barad shows, the point that Bohr is making is much more radical: it raises a profound ontological question as opposed to an epistemological and methodical one – a question about the nature of reality, not merely our knowledge of it and the limits of that knowledge. In doing so, Bohr questions the entire history of Western thought as well as physics as practiced since Newton, based on the view that the world is populated with *individual things*, being what they are in separation and isolation from each other and each possessing their own independent sets of determinate properties of such a nature as to allow for exhaustive knowledge. While Heisenberg rightly pointed to the importance of measurement in our quest for knowledge, Bohr recognized that its full implications play out on a much more constitutive level. Barad puts it like this: "There is something fundamental about the nature of measurement interactions such that, given a particular measuring apparatus, certain properties *become determinate*, while others are specifically excluded. Which properties become determinate is not governed by the desires or will of the experimenter but rather by the specificity of the experimental apparatus" (19). Again, we see how radical Bohr is: like all other classical properties – such as an individual entity's position and momentum in Newtonian physics – an intentional state of mind, assumed to exist in a separate individual human agent (researcher), and that cannot be taken for granted. Intentions, that is to say, are not preexisting determinate mental states of individual human beings. Anticipating the claims she will make on behalf of her agential realism, Barad is prepared to take her reading of Bohr into territories that he barely entered, suggesting that "perhaps intentionality might be better understood as attributable to a complex network of human and nonhuman agents, including historically specific sets of material conditions that exceed the traditional notion of the individual. Or perhaps it is less that there is an assemblage of agents than there is an entangled state of agencies" (23).

What Barad considers Bohr to have helped us appreciate as a consequence of quantum physics is that "we are a part of that nature that we seek to understand" (26), a somewhat commonplace way of putting a claim on what Barad wants to develop in a radical fashion, as we will see. Any attempt to understand the world must take into account the fact that our knowledge-making practices are what Barad calls "social material enactments" that contribute to, and are a part of, the phenomena that we describe. While Bohr, as we will see, stopped short of following through on the full consequences of his insights, he did pave the way for the critique of *representationalism* that was to become crucial in various strands of late twentieth-century thinking, spanning poststructuralism, deconstruction, and feminism. Bohr objected to the idea that knowledge is about the way in which one separately existing entity, the subject, tries to access another such, the object; whereby both entities remain exactly like they were prior to the process, the subject taken as the active party, the object as the passive one, being what it is and allowing the knowledge-seeking subject to mentally copy its innate individual properties by way of "representations" in the mind. The epistemology of representationalism harbors a series of ontological dualisms whose validity in assumedly mind-independent reality is simply taken for granted: subject versus object, mind versus matter, phenomena versus noumena (Kant), inner versus outer, culture versus nature, and the social versus the natural.

What is remarkable about Bohr's contribution to this critique is his willingness to question representationalism's distinction between words and things (objects). Not content to highlight *either* the discursive *or* material nature of practices, Bohr was keen to take both into account at the same time, regarding them as co-constitutive of each other, that is to say, as what Barad depicts as *entangled*. To be entangled, Barad tells us in her book's first sentence, "is not simply to be intertwined with another, as in the joining of separate entities, but to lack an independent, self-contained existence" (ix). Individuals, then, do not preexist their interactions; instead, they emerge through them. Now, while Bohr-inspired Barad to develop her notion of entanglement, and while he clearly broke with the ontology of Newtonian physics and the representationalist epistemology it subscribes to, she nonetheless finds that Bohr "stays singularly focused on the epistemological issues" (31). Discarding the traditional notions about independent objects and their innate properties, as well as the correspondent premise of a self-contained stable epistemic subject, Bohr recommended that we talk about "phenomena". Barad takes this further in seeking to elaborate its full ontological as distinct from merely epistemological implications: in her agential realist elaboration, phenomena do not merely mark the epistemological inseparability of observer and observed that Bohr saw as following from Heisenberg's principle of uncertainty; rather, phenomena according to Barad are "the *ontological* inseparability of agentially intra-acting components" (33).

To appreciate the ontological depth of Barad's Bohr-inspired agential realism, we need to look more closely at the philosophical core of the disagreement between Heisenberg and Bohr. In retrospect we may say that Heisenberg's principle of uncertainty dealt a severe – many would say lethal – blow to the hegemony of representationalist epistemologies: that which is represented could no longer be held to be independent of all practices of representing. The "disturbance" caused to the object through the measurement enacted by the knower that so worried Heisenberg, and that he sought for ways to circumvent and neutralize so as to get at the object in a perfectly direct and in that sense objective way, to be repeated, in principle, by an endless series of separate "knowers" and replicated in observations and experiments so as to yield the same result at all times and places, is nothing of the sort: it is productive and inevitable; it sets up the very context within which such a thing as knowledge is possible, if at all. As Joseph Rouse points out, the representationalism that is operative in Heisenberg's worry is a byproduct of Cartesianism: it has to do with the asymmetrical faith in *word over world* that underlines the nature of Cartesian doubt: "The presumption that we can know what we mean, or what our verbal performances say, more readily than we can know the objects those sayings are about, is a Cartesian legacy, a linguistic variation on Descartes' insistence that we have a direct and privileged access to the contents of our thoughts which we lack towards the 'external' world" (Rouse 1996: 209; Barad 2007: 49).

However, the Cartesian privileging of our access to representations over things has not been without detractors. Giambattisto Vico pioneered the rival view that in order to gain knowledge, the knower has to intervene into the object to be known: only by impacting on it, directly and physically, will knowledge be forthcoming. Strangely ignored in Barad's discussion, Vico's view anticipated what is nowadays termed the *performative* approach, though it must be said that Vico was not so radical as to abandon the premise of an elementary and lasting division between knower and known. What he did recognize, however, and what resonates with Barad's project, is that "knowing does not come from standing at a distance and representing but rather from a direct engagement with the world" (Barad 2007: 49).

Historically, to reject representationalism has often been associated with relativism in that the referent to establish objectivity and validity is seen to disappear, its place left vacant, as it were. Interestingly, Bohr, and, in her own way, Barad, argue to the opposite effect, supported by the argument given by philosopher of science Ian Hacking that manipulability – clearly a case of performativity, taken here as the ability to intervene effectively – is nothing less than "the criterion for determining what is real": "Experimenting on an entity", maintains Hacking, "does not commit you to believing that it exists. Only *manipulating* an entity, in order to experiment on something else, need do that". Observing that physicists doing experiments cannot help but be realists toward the entities that they use as tools, Hacking lays out his criterion as follows: "We shall count as real what we can use to intervene in the world to affect something else, or what the world can use to affect us" (Hacking 1983: 146, 263). Far from opening the door to relativism, then, our ability to effectively intervene provides the strongest case for realism, whereby Hacking distinguishes between realism toward entities, so-called "ontological realism", and realism toward theories, so-called "epistemological realism", approving of the former but not the latter. For Hacking, not representing, as in theorizing, but intervening, as enacted in experiments, is the basis for realism.

In making experiments, the paradigm case for how we attempt to gain knowledge, Hacking echoes one of the points made by Thomas Kuhn in his classic *The Structure of Scientific Revolutions*: that observing something is everything but a naive act of which everyone is capable, simply by concentrating and using one's eyes, as it were. The "something" to be observed is not just anything "out there"; it is a highly selective and selected "something", noticed, and understood for what it is, only by someone who has successfully gone through a process of socialization: spotting what the experiment is about, what it is meant and expected to allow to show up, takes a lot of practice. What counts as representations is produced by specific practices: for lack of these, there is nothing to represent, to observe in the sense intended. Representations, therefore, are not snapshots or depictions of what passively awaits the observer but are "condensations or traces of multiple practices of engagement" (Barad 2007: 53). Contrary to Hacking's realism toward entities, but not theories, Bohr, Barad reminds us, developed a proto-performative account showing that theorizing, too, must be understood as an embodied practice, not "a spectator sport of matching linguistic representations to preexisting things". Concepts, according to Bohr, are themselves specific physical arrangements, ways of engaging with what we seek to understand that help co-produce it, make it what it is.

Whereas measurement plays no systematic role in Newtonian physics (besides "recording" an observation-independent reality as it presumably is in itself), Bohr insisted that quantum physics calls for a new framework that recognizes the constitutive role of measurement processes in the construction of knowledge. To understand the more specific claims that Barad advances by way of her use of Bohr, we need to look at the issue that first and foremost helped launch the new physics: light.

II

What is light? Is it a wave? Or is it a particle?

Thomas Young's double-slit experiment is commonly credited with delivering the final blow to Newton's corpuscular or particle theory, providing unambiguous evidence that light is a wave. And in the 1860s, the physicist James Clerk Maxwell worked out a unified field theory of electric and magnetic phenomena, making it possible to derive, rather than merely postulate, the wave nature of light.

Then, during the first quarter of the twentieth century, the situation changed. As Barad notes, it became increasingly difficult to understand how *any* consistent understanding of the nature of light would be possible. Indeed, the new experiments that were carried out suggested the scenario that was to cause Heisenberg so much concern: light manifested particle-like characteristics under one set of experimental conditions and wave-like characteristics under another. Still more perplexing developments were to follow, however, since evidence suggested that *matter* exhibited this same dual "wave-particle" feature. Matter, explains Barad, "could exhibit wave behavior as well as the (classically expected) particle behavior (under complementary circumstances). Wave-particle duality seemed to be a feature of both light and matter" (Barad 2007: 100).

Today, given the influence that quantum physics has exerted not only in science and philosophy but in our entire cultural outlook, it is difficult to realize the shock that these findings caused to the foundations of Newtonian physics, and the sense of intellectual vertigo that followed. Waves and particles were regarded as distinct phenomena with mutually exclusive characteristics exhibited in separate entities. Whereas particles occupy a given location at each moment in time, waves have extension in space, occupying more than one position at each moment in time.

Accordingly, "the dual nature of light and matter presented a quandary of the first order: an object is either localized or extended; it can't be both" (ibid.).

Bohr spent many years pondering the paradox of wave-particle duality. He came to understand that the experiments that displayed this dual nature were both consistent and reproducible, and hence allowed for scientific objectivity, despite the puzzlingly, yet stubbornly paradoxical findings: "every time a given apparatus was used, the same behavior – whether particle or wave (or both) – resulted. One apparatus consistently manifested one kind of behavior, and a mutually exclusive apparatus consistently exhibited another" (Barad 2007: 105f.). Bohr insisted that we have to face the fact that wave and particle behaviors are exhibited under *complementary* – that is, mutually exclusive – circumstances. Depending on the apparatus we are using, what we will obtain will *either* be a pattern that characterizes particles *or* one that exhibits a wave pattern. There is no option to have it both ways at once. In other words, we have to accept as true the conclusion that, to quote Barad, "the nature of the observed phenomenon changes with corresponding changes in the apparatus" (106). The shift involved could be put more strongly still: it is the shift from the *epistemological* point that what you observe will depend on the apparatus that you are using, to the *ontological* one that what what you observe *is* depends on the apparatus used in the sense of being partly produced and constituted by it.

By his willingness to admit the truth of this conclusion, immensely provocative at the time, Bohr explicitly called into question the fundamental assumptions about methodology, epistemology, and ontology that Newtonian physics in general and Newtonian measurement practices in particular had relied on – assumptions that for centuries had been regarded and acted upon as the bedrock of science *per se*. As Barad observes (and I paraphrase; see Barad 2007: 107), there are three major assumptions involved, all challenged to the core by Bohr's acceptance of the notion of complementarity: first, the belief in representationalism understood as the independently determinate existence of words and things; second, the metaphysics of individualism in the sense that the world is composed of individual entities with individually determinate boundaries and properties; and third, the intrinsic separability of knower and known in that measurements reveal the preexisting values of the properties of independently existing objects as separate from the measuring agencies and the apparatus used. It follows from Bohr's questioning of all three assumptions that the strict determinism that is the hallmark of Newtonian physics is challenged too: the idea, that is, that given the so-called "initial conditions (i.e., the position and momentum of a particle at any one instant in time) and the full set of forces acting on a particle, the particle's entire trajectory (i.e., its entire past and future) is determined" (107). Once the laws of classical mechanics are no longer presupposed as valid, the Newtonian equations based on them will no longer be able to do the job that it was credited with of predicting the physical state of a system for all time.

Readers who at this point get the impression that Newton was a helplessly naive realist, should think twice. The core of the disagreement between Newtonian physics and the new physics is more subtle than that. That things are disturbed when we measure them, and that this is so necessarily and inevitably, is not, as Barad reminds us, a novelty to emerge with quantum physics in the twentieth century – it is a fact acknowledged in the classical view. The reason why Newtonian physics was not worried about the fact of disturbance is that it held that in each case the disturbance could be determined and subtracted out; that is, it could in principle be controlled so that the objectivity and validity of the observations obtained are not threatened. The act of subtraction was conceived as a deliberate and controlled one; it would therefore help secure a measurement-independent grasp of the object's values – i.e., its properties as they are "in themselves", independently of the measurement and apparatus employed in the particular case.

Barad is good at explaining Bohr's reasons for rejecting the Newtonian complacency over the fact of disturbance as intellectually flawed. Specifically, the discontinuity and the indeterminacy of measurement interactions are factors here to stay and not to be controlled in the manner assumed by Newton: the notion of measurement transparency has to be abandoned. The insight into the "essential discontinuity" of measurement interactions, breaking down the strict barrier between knower and known, is indeed one of the hallmarks of quantum physics that marks its revolutionary character. In other words, Newtonian physics will not be able to "ignore measurement interactions by presuming that they can always be reduced to the point where they are negligible". The deeper reason why this is so, highlighted by Bohr, is that the premise on which Newtonianism relies, namely that it will always be possible to determine the effects of the measurement interaction, is false. Bohr's argument for the in principle *indeterminable* nature of measurement interactions is based on his insight that "concepts are defined by the circumstances required for their measurement", which means that "theoretical concepts are not ideational in character; they are specific physical arrangements" (Barad 2007: 109). Note that this is not Bohr's way of putting it; it is Barad's, anticipating a key point in her agential realism. For Bohr as for Barad, however, "measurement and description (the physical and the conceptual) entail each other (not in the weak sense of operationalism but in the sense of their mutual epistemological implication" (ibid.). It follows that to determine all the required features of the measurement interaction (as aspired to in Newtonian physics), mutually exclusive experimental arrangements would need to be employed simultaneously. But as Bohr argues, this is precisely what is ruled out: it cannot be done.

In important respects, the insight that Barad wants to take from Bohr recalls Hans-Georg Gadamer's discussion, in *Truth and Method*, of the so-called "prejudice" that the scholar brings with her as part of a "hermeneutical baggage" when approaching a text and trying to interpret its meaning. In fighting the stubborn "prejudice against prejudice" that he traced back to the rationalist strand of Enlightenment thinking and found still operative in the philosophy of science dominating not only the natural sciences but the social and human ones (*Geisteswissenschaften*) as well, Gadamer argued that approaching one's object of inquiry from some *particular* point of view and with a set of preconceptions about what to encounter in it and to identify as "interesting" and "relevant" about it, does not relativize and so endanger the validity of the findings thus accessed. On the contrary, it is the only possible *and* valid way to engage with *any* object of inquiry so as to formulate hypotheses about it that deserve to be taken seriously by others. Selectivity – *this* set of preconceptions and questions, not just any – does not disturb, let alone destroy the possibility of stating something valid and truthful about one's chosen object, but instead makes doing so possible. Being non-optional and ineluctable, Gadamerian selectivity is precisely *not* a kind of property of the individual subject (scholar) that he or she disposes over and whose impact can be turned on and off. It is therefore not something that should – or that can – be neutralized or bracketed so as to subtract and thus remove its impact from what is forthcoming in the course of the inquiry. It follows that selectivity in the sense elucidated by Gadamer necessarily goes both ways (to allude to the knower/known distinction being debunked here): it involves all the parties and factors making up the undertaking of every concrete inquiry. In short, selectivity in the fundamental sense intended is not something to regret: it is to be welcomed and to be used for all it is worth in each case, as an eminently productive resource, as a *sine qua non* for inquiring into and hopefully gaining knowledge about anything at all – which in practice always means: something specific and limited.

Gadamer is nowhere mentioned in Barad's book but his critique of the assumptions that nourish the notion that the factor of "disturbance" can be controlled so as to make no difference at all and allow us to access reality in a state of purity is clearly instructive for her project. Even so, there is a radical edge to Bohr's most far-reaching insights that goes beyond Gadamer's contribution, and possibly also beyond Thomas Kuhn's demonstration, in *The Structure of Scientific Revolutions*, of the indispensability of looking, not for everything that might "show up", but always for something particular, and to do so in like manner as does the scientific community one belongs to, within the hard natural sciences – including physics – as well as in the social and human ones that Gadamer focused on. We need to probe more deeply to appreciate what this radicalness consists in.

III

Taking the impossibility of determining the proton's position and momentum simultaneously as her example, Barad investigates the consequences of Bohr's thesis that "observation is only possible on the condition that the effect of the measurement is indeterminable". The importance of this, indicated above, is that "we can't subtract the effect of the measurement and thereby deduce the properties that the particle (is presumed to have) had before the measurement" (Barad 2007: 113). Does this mean that we have to give up on measuring position accurately? No, argues Barad, and following Bohr, we can still do that. What it does mean, however, is that "we are *not* entitled to ascribe the value that we obtained for the position to some abstract notion of a measurement-independent object" (ibid.). This being so, the question becomes: what does the value correspond to, what is the objective referent? Is there such a thing, given Bohr's analysis?

Barad explains the consequence that Bohr himself drew as follows: "Since observations involve an indeterminable discontinuous interaction, as a matter of principle, there is no unambiguous way to differentiate between the 'object' and the 'agencies of observation'. No inherent/Cartesian subject-object distinction exists" (114). Depending on the arrangement we set up to facilitate observation, in one case the light (photon) will be part of the agencies of observation; in another case – varying the arrangement – the light's (photon's) momentum is being measured, meaning that it is now part of the object in question. What this variance goes to show is that the question of what constitutes the object of measurement is not fixed: it does not allow for one answer only, excluding all alternatives. Barad's way of expressing Bohr's insight here is that "there is no inherently determinate Cartesian cut: the boundary between the 'object of observation' and the 'agencies of observation' is indeterminate in the absence of a specific physical arrangement of the apparatus". The apparatus, Barad makes clear, "enacts a cut delineating the object from the agencies of observation". She adds the further claim of Bohr's that "the measurement interaction can be accounted for only if the measuring device is itself treated as an object, defying its purpose as a measuring instrument" (ibid.).

Is the radical conclusion that we have no choice but to accept that we must simply give up on the deep-seated request to make a robust distinction between what goes into facilitating that observations be made, on the one hand, and what is being observed – what those observations are about and are focused on – on the other? If there is, and will always be, an oscillation between what counts as agencies of observation, on the one hand, and what figures as the object, on the other, such that what constitutes the one and what makes up the other will vary from case to case, how can you avoid wholesale relativism and arbitrariness, rendering objective scientific knowledge impossible? Or perhaps raising the spectre of relativism is too old-fashioned a way to respond to the new situation. Is not the fundamental issue thrown up here ontological rather than epistemological – ontological in the sense that the dissolution of the division between knower and known, observer and observed, seems to result in reality as such becoming slippery and so eluding any effort that we may make to grasp it, unambiguously and in its totality?

Barad does not engage with these questions directly. As we shall see, however, she eventually does suggest some answers, based to a large extent on the concept she introduced in the above quotation: that of a "cut" or "cut-off point". What is this? And does it allow for the attainment of knowledge that we are after? If so, how exactly is that knowledge to be validated, given the revolutionary implications of the new physics that Bohr – differing not only from Heisenberg but, as we will see, from Einstein as well – never shied away from confronting head-on?

In the first systematic statement about her notion of the cut, Barad writes:

> The specification of the conditions necessary for an unambiguous account of quantum phenomena is tantamount to the introduction of a constructed, agentially enacted, materially conditioned and embodied, contingent Bohrian cut between an object and the agencies of observation. That is, although no inherent distinction exists, every measurement involves a particular choice of

> apparatus, providing the conditions necessary to give meaning to a particular set of variables, at the exclusion of other essential variables, thereby placing a particular embodied cut delineating the object from the agencies of observation. So for every given apparatus, there is an unambiguous resolution of the distinction between the object and the agencies of observation.
>
> (115)

In making the case for the importance of the cut thus conceived, Barad returns to the discussion between Heisenberg and Bohr following the former's famous 1927 paper on uncertainty relations. Contrary to the still quite common view that Heisenberg's thoughts on uncertainty and Bohr's on complementarity are the cornerstones of the so-called Copenhagen interpretation, Barad contends that the two contributions "constitute fundamentally different, indeed arguably incompatible, interpretative positions" (ibid.).

As we saw above, Heisenberg's analysis of the fact that the electron's momentum is "disturbed" by the photon in the attempt to determine the electron's position led him to the conclusion that the uncertainty relation is an *epistemic* matter, highlighting the unwelcome fact, reluctantly admitted, that there is a limitation to what we can *know*. Significantly, Heisenberg does not give up on the notion of a determinate value of the electron's momentum: he maintains the Newtonian belief that such a value *exists* independently of measurement; what is frustrating, however, is that we cannot access it as such and observe it in its purity.

But is it really impossible to determine the effect of the measurement interaction and subtract its effect? As Barad explains, trying to answer this question is what led Bohr to ponder the issues raised by Heisenberg's uncertainty relations more deeply than did Heisenberg – namely by shifting the emphasis from the epistemic/epistemological terrain to the ontological one. It should not be lost on us that in making this move, Bohr's project is reminiscent of Whitehead's, though they approached the ontological terrain from very different angles.

While Heisenberg concentrated almost exclusively on the problems (as he saw it) caused by the discovery of discontinuity as entailed in measurement interactions, Bohr responded by taking a more constructive approach: for Bohr, Heisenberg had helped open up a whole new field of discovery and novel insights: a chance to be exploited intellectually for all it's worth rather than regretted in that it necessitated throwing overboard a number of old-worn certainties. Specifically, Bohr focused his attention on the conditions of possibility (Kant) for determining the effects of the measurement interaction. Bohr realized that "concepts are meaningful, that is, semantically determinate, not in the abstract but by virtue of their embodiment in the physical arrangement of the apparatus" (Barad 2007: 117). Heisenberg's analysis stopped short of addressing the question of how what Barad calls the cut – delineating the object from the agencies of observation – gets made: the very question that needs to be answered if the indeterminacy is to be resolved. For Bohr, that is, the deeper issue is precisely that of indeterminacy, not uncertainty. He proposes that "the reciprocal relation between position and momentum be understood in semantic and ontic terms" (the terms are Barad's; see 2007: 118), and only derivatively in Heisenberg's epistemic ones.

The fundamental issue with regard to the nature of measurement, then, has not to do with disturbance and how to control it (Heisenberg would prefer removing it, if possible) but with resolving an inherent indeterminacy. The key point for Bohr, according to Barad, is what he termed "quantum wholeness", that is to say, the lack of an inherent (Newtonian and Cartesian) distinction between the object and the agencies of observation. There being no unambiguous way to make the cut – to differentiate – between the two, an apparatus must be employed to resolve the ambiguity, whereby "the apparatus must be understood as part of what is being described" (ibid.). "Quantum wholeness", the key notion Bohr suggested here, is closely connected to another one, "phenomenon", used to designate particular instances of wholeness. In a 1958 essay, Bohr explains what is at issue: "While, within the scope of classical physics, the interaction between object and apparatus can be neglected or, if necessary, compensated for, in quantum physics this interaction thus forms an inseparable part of the phenomenon. Accordingly, the unambiguous account of proper quantum phenomena must, in principle, include a description of all relevant features of the experimental arrangement" (Bohr 1958: 4; Barad 2007: 119). What Barad terms "the Bohrian cut", then, "marks off and is part of a particular instance of wholeness, that is, a particular phenomenon" (ibid.).

Importantly, and perhaps surprisingly, Bohr insists that quantum mechanical measurements are "objective". Yet Barad is quick to point out that Bohr "cannot possibly mean that the measurements reveal 'objective' (i.e., premeasurement) properties of independent objects" (119); this of course follows from his critique of Newtonian physics. What, then, does Bohr mean by "objectivity"? Barad quotes his remark that "no explicit reference is made to any individual observer" (ibid.). "Objective" means reproducible and unambiguously communicable, in the sense that "permanent marks [are] left on bodies which define the experimental conditions" (ibid.). Now, while Barad maintains that Bohr's notion of objectivity is not predicated on an inherent or Cartesian/Newtonian distinction between objects and agencies of observation, the statement of his she quotes to this effect is in fact wholly conventional – namely that "we can by an objective description only understand a communication of experience which does not admit of ambiguity as regards the perception of such communication" (Bohr quoted in Barad 2007: 120). What is clear is that Bohr sees no threat of relativism looming, in the sense that one communicator will make observations/statements that cannot be validated by others. But the deeper, ontological question is not addressed: that of how the experience to be communicated is accessed in the first place.

To offer a fuller account of Bohr's position, Barad brings together his notion of the physical apparatus and his notion of concept, so that "the physical and conceptual apparatuses form a nondualistic whole marking the subject-object boundary" (ibid.). In keeping with Bohr's aim to concretize concepts by way of situating them, applying them as one factor among several in what he calls the "phenomenon of wholeness", concepts attain their meaning in relation to – in a specific arrangement with – a particular physical apparatus. According to Barad, this is what "marks the placement of a Bohrian cut between the object and the agencies of observation, resolving the semantic-ontic indeterminacy". This resolution "provides the condition for the possibility of objectivity" in Bohr's account, in the sense that objectivity requires accountability to what Bohr refers to as "permanent marks" – "such as a spot on an photographic plate, caused by the impact of an electron – left on the bodies which define the experimental conditions" (Bohr quoted in Barad 2007: 120).

But this still leaves unanswered the question: what is the objective referent for the determinate value of the property measured? On the reading of Bohr that Barad proposes, measured properties refer to phenomena, whereby phenomena in Bohr's understanding include "all relevant features of the experimental arrangement", including, as we just saw, the physical apparatus as well as the concepts applied. Thus, and importantly, the referent shifts: it is not the observation-independent object Descartes and Newton presupposed, but a phenomenon. Bohr's point is that, far from jeopardizing objectivity, this shift secures it: it is a condition for the possibility of objective knowledge. On this basis, Bohr proposes that the wave-particle duality paradox that initially had caused so much worry, can be solved as follows: "wave" and "particle" are "classical concepts that are given determinate meanings by different, indeed mutually exclusive, apparatuses and that refer to different, mutually exclusive *phenomena*, not to independent physical objects". Bohr hoped that this solution would save the theory from inconsistencies, "since it was impossible to observe particle and wave behaviors simultaneously because mutually exclusive experimental arrangements are required" (Barad 2007: 121).

In sum, Bohr's most important and far-reaching contribution consists in arguing that the host of seemingly intractable issues thrown up by the uncertainty/indeterminacy problem that he originally discussed with Heisenberg are in essence of an ontological rather than epistemological nature, and that their being ontological does not endanger scientific knowledge and its objectivity but instead furnishes it with an entirely new foundation. Barad's way of summing this up is worthwhile quoting, not least because it says a lot about how she wishes to exploit Bohr's argument for her own systematic purposes:

> The measurement apparatus is the condition of possibility for determinate meaning for the concept in question. [. . .] Apparatuses provide the conditions for the possibility of determinate boundaries and properties of "objects" within phenomena, where "phenomena" are the ontological inseparability of objects and apparatuses. Since individually determinate entities do not exist, measurements do not entail an interaction between separate entities; rather, determinate entities emerge from their intra-action. [. . .] a phenomenon is a specific intra-action of an "object" and the "measuring agencies". [. . .] We should understand phenomena not as objects-in-themselves, or as perceived objects (in the Kantian or phenomenological sense), but as specific intra-actions. Because the basis of this ontology is a fundamental separability, it cuts across any Kantian noumena-phenomena distinction: there are no determinately bounded or propertied entities existing "behind" or as the causes of phenomena.
>
> (128)

IV

Turn now to Barad's case for "agential realism". Having laid out the nature of her debt to Bohr, it is time to explore what is truly novel in Barad's project. Doing so will allow us to appreciate Barad's distinct approach to some of the core issues discussed in the previous chapters on Nagel and Whitehead, in particular the nature of matter, the place and dimensions of agency, and the question of how to assess panpsychism. It will also open up the field of issues common to various versions of posthumanism, as we will see in due course.

Crucial to Barad's project is the notion of a "cut", a notion profoundly inspired by Bohr yet going beyond his understanding of it. As we just saw, individuation – what counts as an "object" – as conceived by Bohr is not a given and is not something ontologically stable, in the Kantian sense of possessing a set of in-itself properties that, presumably, are what they are prior to and independently of any attempt to get at them and describe them. Instead, individuation is the result of specific "cuts" enacted by the experimental arrangement; it will therefore vary from one such to the other. "Agential separability" – Barad's term, not Bohr's – is what establishes the ontological condition for objectivity, understood as "an agentially enacted ontological separability within the phenomenon" (175). Since the agential cut occurs *within* the phenomenon, changing the apparatus will yield a corresponding change in the agential cut and hence in the delineation of objects from agencies of observation and so also "in the causal structure (and hence the possibilities for 'the future behavior the system') enacted by the cut. Different agential cuts produce different phenomena" (ibid.).

Agential realism as elaborated by Barad debunks established notions of agency and proposes that agency must be radically rethought, especially with regard to its place within matter. In saying "within matter", I deliberately allude to the kinship that emerges between Barad and Whitehead, even though references to his work are completely – and conspicuously – absent in her book.

Before we can raise the question of panpsychism in Barad, however, we need to take a closer look at her conception of agency. To "cut agency loose from its traditional humanist orbit", from its alignment with human intentionality or subjectivity, is one of the stated aims of Barad's agential realism (Barad 2007: 177). What she is after is expanding the space of agency, rendering it substantially larger than what is allowed for in such tradition-provoking posthumanist thinkers as Michel Foucault and Judith Butler. As we might expect to follow from this, "the space of agency is not restricted to the possibilities for human action". But neither, adds Barad, is it "simply the case that agency should be granted to nonhumans as well as humans, or that agency can be distributed over nonhuman and human forms". If not, then what? Barad's answer is that "what is at issue, rather, are the possibilities for the iterative reconfiguring of the materiality of human, nonhuman, cyborgerian, and other such forms" (Barad 2007: 178). As conceived in agential realism, then, agency is "a matter of intra-acting; it is an enactment, not something that someone or something has. It cannot be designated as an attribute of subjects or objects (as they do not preexist as such)" (ibid.). Not a matter of what a fixed somebody (subject) intends or initiates, agency is instead "doing" or "being" in its intra-activity: it is "the enactment of iterative changes to particular practices – iterative reconfigurings of topological manifolds of spacetimematter relations – through the dynamics of intra-activity" (ibid.).

Having set out her understanding of agency, Barad proceeds to connect it with her earlier account of the cut. In doing so, she breaks new ground as compared with Bohr's contribution. While highly stimulating, this is also, I shall argue, the point where Barad's project becomes increasingly vague, making claims so all-inclusive and global in scope as to overgeneralize from the limited pool of examples, observations, and evidence meant to corroborate them. Barad's ambition is to redirect the insights that she adopts from Bohr and the consequences that he drew from the commotion sparked by quantum physics. The danger is that her agential realism overstretches that legacy, losing accuracy and plausibility along the way.

Return to the cut. We have seen that for Barad, different agential cuts materialize different phenomena. Hinting at what this means for an ethics, Barad states that "we are responsible for the cuts that we help enact not because we do the choosing [. . .] but because we are an agential part of the material becoming of the universe. Cuts are agentially enacted not by willful individuals but by the larger material arrangement of which 'we' are a 'part'. [. . .] Intra-actions cut 'things' together and apart. Cuts are not enacted from the outside, nor are they ever enacted once and for all" (Barad 2007: 178f.). To be sure, it is difficult not to concur when Barad writes that "our intra-actions contribute to the differential mattering of the world" (Barad 2007: 178). But in speaking of "our" intra-actions to explain the claims she makes on behalf of her agential realism, she invokes the human subject – be it in the first person plural rather than singular – in a way that suggests what she otherwise forbids herself as a self-proclaimed posthumanist: taking the human point of view, or human agency, as the privileged vantage point. True, she may respond that her doing so is due to heuristic or pedagogic reasons, not substantial ones – witness the embarrassment betrayed in her use of scare quotes in the passage just quoted, invoking an unaccounted-for "we".

The fact is that Barad's book is redundant with passages that make the point that the cut delineates "object" from "agencies of observation". She reiterates and affirms again and again Bohr's claim that objectivity is not predicated on an inherent or Cartesian cut between "observer" and "observed"; rather, what is required for objectivity is "an unambiguous and reproducible account of marks on bodies" (Barad 2007: 320 et passim). Individuation, we are told, is "the result of specific intra-actions that entail the larger material arrangement". Barad suggests "a strengthening of Bohr's epistemic notion of objectivity, replacing it with an ontological conception that foregrounds the issue of ontological separability, as Einstein would have it, but without grounding it in the problematic criterion of spatial separability". And she continues by observing that "it has become increasingly difficult in recent years to hold to a belief in the separability condition", thus contradicting flatly, as did Bohr before her, Einstein's despairing comment in a letter to Max Born in 1948: "If one renounces the assumption that what is present in different parts of space has an independent, real existence, then I do not at all see what physics is supposed to describe" (Einstein quoted in Barad 2007: 319). Indeed, as Isabelle Stengers observes in *The Invention of Modern Science*, "it was when Einstein realized he could not construct an internal critique of quantum mechanics that he suggested condemning it in the name of the hope, which identifies the physicist, of constructing an objective representation of the world, independent of observation" (Stengers 2000: 174).

Let us take stock. There is no doubt that Barad's book makes a number of highly important and interesting points about fundamental issues in twentieth-century philosophy of science, in epistemology, and in ontology. The way she goes about actualizing and further elaborating Bohr's contribution – long neglected among the majority of philosophers – is original and thought provoking. In many respects, Bohr was ahead of his time. Take performativity. Drawing on Bohr, Barad shows that there is no disengaged way of getting at what is physical reality. The spectator model as well as the contemplative or introspective one (Descartes), far from helping us gain the access we are after, are in fact non-starters. Only by actively affirming what Heisenberg, and in a somewhat different manner, Einstein, never stopped seeing as a source of worry, as threatening the whole epistemic project of physics, that we may make progress as far as knowledge about the world is concerned: by embracing the impact yielded by factors of "disturbance", realizing that were it not for those factors, as operative in measurement and in the application of the (meaning any) apparatus, we would have no access whatsoever to the domain of so-called epistemic objects. We are part of the nature that we seek to understand. We must engage with what we seek to know. Stronger, we must actively mingle with it. Stronger still (and this is Barad stretching Bohr), we are always already entangled with it, never quite separate, or fully separated, from it. We are in the thick of it, whoever or whatever we are; and whatever "it" is, it is in the thick of us. Meaning is entangled with matter, matter with meaning.

Now take selectivity. There is a limit to what we can get at simultaneously. We always have to select. Rather than observing, disinterestedly and from outside, a separate object with inherent properties, we are engaging with a "phenomenon" such as is partly constituted, made into what it is, by our accessing it by way of an apparatus, by applying concepts, indeed, by partaking in a unique experimental arrangement to the effect that the long-standing dualisms between subject and object, knower and known, thing-for-us and thing-in-itself, interiority and exteriority, mind and matter breaks down. Again, the divisions doing so do not jeopardize objectivity and the gaining of valid knowledge but make it possible. Since apparatus, concepts, and experimental arrangement are active – agential – in co-producing what is to be seen, what emerges to catch our attention, for want of them there will be nothing to access at all. There is no reality as such, meaning independently existing and simply – self-sufficiently – being what it is. Selectivity is thoroughgoing and pervasive in that it is always a determining factor in what we are getting at – and we can never get at it all at the same time, in one strike. Something that is illuminated, is what *this* particular constellation of impacting factors combines to facilitate emerging. Everything else is not exactly left in the dark (as the epistemological take would have it) but is lying dormant, as it were: qua nonactivated, qua nonparty to any performed specific arrangement and entanglement, it is to be reckoned as part of "reality" only negatively or by default – a potential piece of reality not an actual(ized) one.

Move now to apparatus. There is Maurice Merleau-Ponty's famous example with the blind man and his stick. How does the cut work here? Taken as an apparatus, the stick is "productive of the phenomenon it measures" (Barad 2007: 334). So, is the stick on the side of the subject or object, the knower or the known. Well, it permits the blind man to orient himself in the world; better, to have a world, not just any one but *his* world, here and now. It is better still, to produce the whole, yet unique constellation in play at any time and place: man-stick-world so that for lack of the stick as mediating the former and the latter, they would both fail to prevail. To ask where the man ends and where the stick begins, or whether the stick belongs to the world or to the man, to something on the subject-side or something on the object-side, is to ask the wrong questions: their lack of unambiguous answers is not an embarrassment; if anything, posing the questions is (to allude to Heidegger's famous quip about asking, and asking anew, for a knock-down proof of the existence of the physical world).

So far, so good. But how far, and how good, exactly? As the many quotations amply show, from the very first sentence to the last Barad's rhetoric is global, self-consciously and declaredly so. The first: "This book is about entanglements" (Barad 2007: ix). The last: "We need to meet the universe halfway, to take responsibility for the role that we play in the world's differential becoming" (Barad 2007: 396). The programmatic statements made on behalf of her agential realism suggest that we have to do with what I call a "global" account of how knowledge comes about: not just in one particular case, with one particular sort of "knower" and "object", but in all of them, allowing for Bohr's insight into the constitutive role that the constellation of apparatus, measurement, concepts and experimental arrangement will have in each and every case of "producing" the "phenomenon" as explained above.

On my view, while global in reach and ambition, Barad's agential realism is carried out in a way that threatens to weaken rather than validate its stated aims. Attesting to my "global" reading is Barad's oft-repeated insistence that, in her "posthumanist" account, "meaning is not a human-based notion", intelligence is not a specifically human capacity, measurement is not an irreducibly human-centered notion, and apparatus, as the stick example shows, is nothing separate or external from the "knower", and so on (see Barad 2007: 335). Insofar as her posthumanism is meant to identify and critique the limits and blind spots of anthropocentrism, that's fine with me. At this stage, that is not the problem.

The problem is that physics is, and throughout remains, Barad's model: it provides the framework for all her theoretical claims and the material for her illustrations and empirical examples. Physics is a natural science. It relies on measurement, apparatus, and experiments, laboratory such, standardized and repeatable such, allowing for the reproducibility of its results that Bohr – standing tall when both Heisenberg and Einstein suffered a failure of nerve – was so keen on stressing so as to save objectivity in the wake of the controversies caused by quantum physics. And yet, for all his boldness in embracing the revolutionary consequences of the new paradigm in physics, there is something perfectly conventional about the notion of objectivity that Bohr subscribes to. Barad fails to take sufficient note of this and to critically discuss it.

Let me explain what is at issue here. In holding objectivity to depend on the possibility of the repeatability of the experiment and its results – setting it up so as to enable looking for the same thing will yield the same result, regardless of how many times it is done and who does it – and in concluding that these very conditions and criteria for "doing science" still obtain, no matter how revolutionary quantum's physics' break with Newtonian physics might be, Bohr commits himself to a far from tension-free balance between the old and the new paradigm. In saying this, I have not forgotten how radical Barad presents Bohr as being when it comes to rethinking the notions of phenomenon, apparatus, and concept. Nevertheless, the bottom line is the same as it used to be prior to the revolution: sameness is what is required to secure objectivity. So relax, do not lose your nerve and throw the baby out with the bath water: sameness obtains and persists in doing so. Even under radically novel circumstances, the phenomenon both must be, and in fact is, repeatable from one instantiation to the other, showing up the same results for everyone who is part of the arrangement and curious about what will emerge, based on a shared understanding about what is going on and what to look for: never just anything, always something particular (to allude to Gadamer and Kuhn). Bohr's proposed novel way of appreciating the active (Barad: agential) role played by the apparatus in the production of the total phenomenon (always a multi-factor constellation) does not alter the essential basis and criteria for objectivity: the phenomenon must be, and will be, repeatable from one case to another. In short, the traditional formula for the laboratory experiment – "whenever X, then Y" – need not and should not be given up: it still applies, still obtains. True, in one case of setting up the experiment, the result will be "wave", in another, "particle". We recall that Bohr's principle of complementarity is all about it being either the one or the other, never both at the same time: if an experimental arrangement is suitable for defining "position" (e.g. fixed parts), then it is not suitable for defining the complementary variable, "momentum" (e.g., movable parts). Complementarity as understood by Bohr, Barad stresses, is "a matter of material incompatibility: the instrument cannot have a part that is simultaneously both movable and fixed". It follows that "the only well-defined (unambiguous) concepts that one has available are the particular concepts embodies in the specific experimental arrangement. There are no others" (Barad 2007: 329). Now, the fact that the apparatus is not "outside" the phenomenon, but inside it, part of it, is fully compatible with objectivity. Whereas Heisenberg misunderstood the nature of the problem and drew the wrong conclusion, Einstein remained conservative in the wake of the scientific revolution his theory of relativity had helped launch. In their various ways, they both resisted Bohr's assertion that the apparatus is not a disturbance and does not pose a limit to what we can observe and hope to know; rather, it is part of the multi-factor condition of possibility of knowledge, not what spoils it but what helps facilitate it.

Whereas Einstein, according to Barad, was wrong to view separability, understood as an inherent cut between observer and observe, as required for objectivity, Bohr was right in holding that what is required is "an unambiguous and reproducible account of marks on bodies" (Barad 2007: 339). As we have seen Barad argue, this is made possible by the intra-active enactment of a "cut" that "unambiguously differentiates the 'object' (that which 'causes' the mark) from the 'agencies of observation' (the 'effect' or that which receives the mark)" (ibid.). Her frequent use of scare quotes is notable here. What does it signal? I think it signals that, far from necessitating that conventional concepts such as cause and effect, object and observation, simply be given up, Barad's agential realism aims to liberate them from their epistemological innocence and their ontological purity – their assumed pristine self-containedness – to accommodate them to the new framework she is proposing.

V

This is the point in her argument where Barad's *posthumanism* comes to the fore. Despite their differences and disagreements, Bohr and Einstein share "the humanist belief in the existence of individual humans that are separately determinate from the physical interactions being investigated". By contrast, Barad's posthumanist position does not presume the human to be "a special system separate from the natural processes that he or she observes"; rather, the emergence of the "human" must be understood along with all other physical systems (Barad 2007: 339). The human cannot be excluded from the realm of nature, nor can humans be sequestered in the realm of culture; the traditional nature/culture division as such must be abandoned since the ontological dualism that it relies on – i.e., that nature and culture constitute two different and independent regions of reality – is false.

In rejecting the humanist and anthropocentric bias of both Bohr and Einstein, Barad offers what must be considered her most important criticism of Bohr. The limitations she sees in Bohr's otherwise invaluable contribution to her own project come in two parts. The first has to do with his failure to fully articulate the nature of the "apparatus". Bohr, that is, tends to treat the apparatus as "an ideal measuring device that springs full blown from the head of Zeus"; its constitutionality remains something constant, "frozen at a moment in time, denied its historicity and mutability"; in effect, then, the Bohrian apparatus is "hermetically sealed off from any and all 'outside' influences" (Barad 2007: 144). In this manner, Bohr mistakes the apparatus for a mere laboratory setup, artificially cutting it off from all the "extra-scientific" activities that enable experimental practice to work, as demonstrated in the work of Latour and Hacking.

The second criticism concerns Bohr's understanding of the objectivity-facilitating criteria of reproducibility and communicability, described above. Referring to these criteria, Barad complains that in Bohr's account "the human is cemented into the very foundations of quantum theory". In making "observation and communication" crucial, Barad sees Bohr as holding that "man isn't merely the measure of all things; man's finitude is implicated in the very conditions of possibility of measurability and determinability", concluding that Bohr "overshoots his mark and places the human not merely back in the picture where he or she belongs, but at the center of all that is" (Barad 2007: 143).

The posthumanist stance Barad that promotes, then, refuses to presume that humans occupy a privileged position in physical theories, be it in the form of "a special position inside or outside the realm of natural phenomena or the theory that accounts for them" (Barad 2007: 323). The question that we should ask is: "Should not the 'human' be accounted for in terms of the theory and the specific intra-actions from which it emerges, rather than the other way around?" Having said this, Barad concedes that Bohr can be considered to share this posthumanist stance when he remarked that we are a part of the nature that we seek to understand. At the end of the day, however, Barad finds that what Bohr "offers with one hand, he takes away with the other" (ibid.).

What, more precisely, is Barad's posthumanist agential realism aiming for? What does its alternative ontology to that of humanism and anthropocentrism consist in, seeking, we are told, to understand the emergence of the "human" along with all other physical systems? What is meant by the contention that "intra-actions are not the result of human interventions; rather, 'humans' themselves emerge through specific intra-actions" (Barad 2007: 352)? If so, who are the other-than-human agents that take part in enacting the intra-actions through which humans emerge? Likewise, with regard to the agential cut on which Barad plays such strong emphasis throughout: What does it take to make the cut? Do agents other than human ones take part in carrying it out? If so, who, and of what kind, are they? And is enacting the cut to be understood as something that happens – and need to happen – *deliberately*, as Barad's usage of the term and the examples she gives seem to indicate? If so, would that not place a serious restriction on who the agents in question can possibly be? And last but not least: perhaps these questions are hopelessly old-fashioned, wedded to anthropocentric categories and so altogether unhelpful in recognizing the novelty of what Barad's project is hoping to accomplish. Perhaps, with humankind having entered into the Anthropocene, we need to follow suit philosophically, as it were, by way of upping the ante – and ask: could it be that an altogether *novel sort of agent* is enacting – or effecting – the cut these days, namely global warming? Or the Earth system? Would that not be proof of a truly novel situation, crowded with "hyperobjects" (Morton), calling for a posthumanist paradigm?

I do not see that Barad answers these questions, though she is good at raising them in a thought-provoking manner. To be sure, to focus on what light Barad's proposed agential realism throws on an alternative account of the "human" and its emergence to that offered in anthropocentrism, may well be said to commit the mistake of reclaiming the primacy of the human, at the very moment when the necessity of giving up that privileged position is the thesis being advanced. (I shall return to this question in Chapter 5 when discussing Latour.)

That said, it is easy to agree with the sort of criticisms Barad is making of Bohr, taking him as an example of how difficult it proves to accomplish the break with anthropocentrism that his realization that we are part of – not apart from – what the nature that we seek to understand calls for. However, Barad's often-made statements about "humans" being "themselves specific parts of the world's ongoing reconfiguring", that is to say, part of "the world-body space in its dynamic structuration", and about knowing being a part of being and not "of the world from above or outside" it (Barad 2007: 341), only take us so far.

As I pointed out above, Barad tends to overstretch her use of Bohr and the applicability of his thoughts on the "new physics" to her own agential realism. This tendency is borne out in Barad's practice of continuing to invoke and make use of Bohr's framework – that of physics – even when she is topically departing from it to elaborate what I call the global reach of her agential realism.

Consider, again, the issue of objectivity. Barad points out, as we would expect, that objectivity cannot be "defined in reference to a human observer: it is not merely about what humans can do to facilitate unambiguous communication about laboratory results" (Barad 2007: 340). That is fine and in line with her announced posthumanist stance. But now witness how she continues the paragraph: "Rather, objectivity is a matter of accountability to marks on bodies", backing this up by referring to how "the agential cut" helps facilitate the "reproducibility and unambiguous communication of laboratory results" (Barad 2007: 340). This is all taken from a passage where Barad explicitly tells her readers that she is explaining her "agential realist elaboration of Bohr's framework" (Barad 2007: 339).

I want to make two points here. First, there is the admittedly trite observation that the agents engaged in the communication of laboratory results, as well as in the issue of their objectivity, assumedly are humans and humans only – despite it being Barad, not Bohr, making the statements just quoted. Likewise, it is presumably only human agents who engage in organizing such a thing as a laboratory experiment, making decisions about what kind of apparatus and instruments to use so as to facilitate there being anything – meaning something specific – to "observe", and who, deciding when the experiment is to be declared completed, devote their attention to determining the results that it yielded and to interpreting them so as to assess their significance in terms of "success" or "failure" in the advancement of knowledge, this being, presumably, their reason for undertaking the experiment in the first place. My point is a simple one: in the very passage where Barad promises her readers a "posthumanist" account, one going beyond the human-centered one that she criticizes in Bohr, she seems to tacitly presuppose specifically human agents – with the capacities, interests, and objectives that go with them – at every point along the way and with regard to all the key concepts being employed. Now it might be said that part of what Barad is after is a more comprehensive understanding of what goes into the "emergence" of such human agents (i.e., scientists, physicists), with special attention given to the extra- and nonhuman agents, factors, and circumstances that help facilitate their existence and their various projects, the carrying out of experiments being one such. In that case, however, we need to hear more about how this happens.

This brings me to my second point, which raises the following question: how representative, and thereupon generalizable, is the laboratory experiment that Barad continues to hold fast to and to invoke again and again, *after* she has announced her criticisms of Bohr's humanist stance and launched her alternative, agential realist one, meant to do justice to the extra-human factors that Bohr neglected? We have already seen how this question comes up in connection with the notion of objectivity. Barad, that is, typically counters Einstein's worry about the prospect for objectivity with the assertion that "there are marks on bodies produced through specific intra-actions"; hence. objectivity is "a matter of exteriority-within-the phenomenon" (Barad 2007: 346). What the so-called agential cut does provide, she maintains, is "a contingent resolution of the ontological inseparability within the phenomenon and hence the conditions for objective description: that is, it enables an unambiguous account of marks on bodies, but only within the particular phenomenon" (ibid.). The solution to the "measurement problem", Barad writes in conclusion to this passage, consists in "recognizing that what is at stake is accountability to marks on bodies in their specificity by attending to how different cuts produce differences that matter" (ibid.).

No further documentation is needed: Barad's way of going about elaborating her alternative stance takes the form of sticking to the framework that was always the showpiece for the one she says she wants to go beyond: objectivity as secured in the setting provided by the laboratory experiment. The question raised by her doing so, yet without offering reflections on it and the tension or even inconsistency it creates for her project, has to do with the role of two constantly invoked features in particular: the "marks on bodies" and the "agential cut". It goes like this: can there, on Barad's view as she presents and defends it, be knowledge such as accessed and gained in another way than through an agential cut enacting marks on bodies? If the answer is that Barad is only trying to throw new light on how a physicist should go about performing an experiment, then that might be fine, and an important thing to do. But it would not offer the reader good reasons to think that her agential realism has the resources and the potential to say something more general, more global, about the mind-blowingly challenging issues that were thrown up by the advent of quantum physics, issues about the nature of nature, about the limits (or lack of such) to human knowledge, and so on and so forth – issues that Einstein, Heisenberg, and Bohr grappled with in their full scientific and philosophical reach, even though they first encountered them as practicing physicists.

So, to put it differently, what if there are no "marks on bodies" to be made, no bodies to impose them on? Is "bodies" just a metaphor in Barad's usage? Is the making and leaving of "marks" also a metaphor? In short, are we talking here about features and factors involved in *all* instances of knowledge-seeking, of the securing of objectivity etc., making up part of the fabric, as it were, in the epistemology and ontology that *always* are in play when knowledge about the nature of the world is sought?

Similar questions must be asked with respect to the "agential cut". Is there always a matter of such a cut being enacted? Is the enterprise of science, or knowledge as such, dependent on it? If not, when is the enactment of the cut a necessary component, and when is it not? If such a distinction applies, why has Barad not discussed it and explained its implications for agential realism? More specifically, if the agential cut is not – as indeed Barad's posthumanism suggests – a prerogative of human agents, that is, if nonhuman such are active in enacting it – presumably in some cases, not in others – then how exactly is the interaction or interplay that in that case obtains between the different agents involved to be understood?

Again, I admit that there may be something excessively pedantic about the way I have been formulating these questions, betraying, moreover, the "old" epistemological and ontological paradigm that Barad is setting out to debunk. And yet, based on what we have seen so far, she has not made good on her promise to clarify her agential realism in greater detail: the aporia that I see is that of elaborating the new within the limited and constrained setting of the old – i.e., the laboratory experiment. We need to find out what precisely her agential realism is meant to achieve, what sort of light if throws on the larger issues of mind, matter, and meaning that Barad announces at the start of her book.

VI

It is first of all in her discussion of Michel Foucault and Judith Butler that Barad addresses the role that materiality plays in agency and performativity. She applauds Butler's plea for a "return to the notion of matter", taken not as a site or surface awaiting human/social/cultural inscription or attribution but as "a process of materialization that stabilizes over time to produce the effect of boundary, fixity, and surface we call matter" (Butler 1993: 9; Barad 2007: 64). Barad also endorses Butler's much-discussed proposal that (human) identity not be understood as an essence but as a doing, and likewise gender not as an essential property of a subject but as a kind of becoming or activity; both claims articulate Butler's rejection of classic humanism's premise about the "givenness of the subject". The upshot is that the question of identity must be linked with the materiality of the body. For all the ingenuity of Butler's criticisms of humanist essentialism, however, Barad considers it "not at all clear that Butler succeeds in bringing the discursive and the material into closer proximity" (Barad 2007: 64).

If questions about the material nature of discursive practices are left hanging in the air in Butler's discussions, they are just as unclarified in those of Foucault, from whose analyses of power Butler draws so much of her inspiration. And even though Barad credits Foucault with the insight that "discursive practices are the material conditions that define what counts as meaningful statements", she nevertheless judges that he is not clear about the material nature of such practices. Foucault fails, that is, to offer an "account of the body's historicity in which its very mentality plays an *active* role in the workings of power". Hence, there is "a reinscription of matter's passivity" in Foucault's work that betrays the remnants of representationalism that haunt his otherwise postrepresentationalist account (Barad 2007: 65). What is more, his notion of the biopolitical field must be viewed as outdated in that it fails to take account of "the new technoscientific practices that continually rework the boundaries between the 'human' and the 'nonhuman'" (Barad 2007: 65), observes Barad, echoing a criticism previously made by Donna Haraway (1997).

The questions Barad finds left unanswered in Foucault and Butler are of the following kind: "If discursive practices constitute a productive social or cultural field, then how much of the very matter of bodies, both human and nonhuman, can be accounted for? Is the matter of things completely social in nature? Are we to understand matter as a purely cultural phenomenon, the end result of human activity?" (Barad 2007: 64).

These are truly big questions, and in raising them in the very general and comprehensive manner she does, Barad risks creating expectations that cannot possibly be met by the agential realism she outlines, meaning it to offer not merely "a unified theory of cultural and natural forces" but to investigate "the very practices through which they are differentiated" (Barad 2007: 66). Rejecting the notion of matter as being an end product of human activity, insisting that matter be seen as "an active factor in further materializations", Barad sets out to do justice to matter "in the fullness of its capacity", thereby offering a robust account of the materialization of *all* bodies, "human" and "nonhuman", including "the agential contributions of all material forces (both "social" and "natural")", hoping to accomplish all of this as part of "a new account of matter's dynamism" as well as "a posthumanist elaboration of the notion of performativity" (Barad 2007: 66). For all the boldness of such a program, there is something breathtaking, and perhaps eventually self-defeating, about committing oneself to such a tall order.

So how does Barad define matter? What is the notion of hers about matter that allows matter to play the active "agential" role that she castigates not only the usual suspects – Cartesian dualists and essentialists old and new – for denying, but fellow critics of humanism like Foucault and Butler as well?

Matter, Barad tells us, is not a passive and blank slate awaiting the active inscriptions of human/social/cultural agents or agencies; like meaning, it is not an individually articulated or static entity, nor is it so many little "bits of nature" or "a fixed support, location, referent, or source of sustainability for discourse"; on the contrary, matter is "always already an ongoing historicity" (Barad 2007: 150f.). While Butler recognizes all these points, being predominantly (and perhaps not accidentally) about what matter is not and how matter has so far been misconstrued, Butler's theory ultimately fails because it "reinscribes matter as a passive product of discursive practices rather than as an active agent participating in the very process of materialization". Much like with the Foucauldian analyses she draws upon, at the end of the day "the processes that matter for [Butler] are only human social practices (thereby reinscribing the very nature-culture dichotomy she wishes to contest)". In contrast, agential realism aims to recognize what Foucault and Butler both miss out on: matter's dynamism, such that "matter is substance in its intra-active becoming – not a thing but a doing, a congealing of agency. Matter is a stabilizing and destabilizing process of iterative intra-activity". Rather than referring to "an inherent, fixed property of abstract, independently existing objects, matter refers to phenomena in their ongoing materialization". This being so, "matter(ing) is a dynamic articulation/configuration of the world" (Barad 2007: 151).

I have several problems with this. For a start, I don't find the way Barad moves back and forth between quantum theory and Bohr's position within it, on the one hand, and Foucault's and Butler's analyses of power, agency, and gender, on the other, helpful in explaining Barad's agential realist view of matter. The jumps made between the different sections and case studies are too abrupt, the argument too loose, leaving her agential realism in a state of philosophical vagueness: radical and suggestive, to be sure, but very hard to pinpoint, to get a firm analytic grasp on. The negatives – what matter is not – never cease to dominate and carry most of the burden of argument; the positives – stressing agency and dynamics, and the entanglement of matter with meaning, and with ethical agency and responsibility – are so many teasers, rarely carried out in any detail and too often left hanging in the air, echoing Barad's own complaint about Butler. For lack of instructive examples and concrete case studies, such as would allow us to know exactly what Barad has in mind when making her sweeping statements about matter and what it does rather than is, it is simply not possible to assess their validity and reach. Asserting in a general fashion that "materiality itself is a factor in materialization", that "matter is enfolded into itself in its ongoing materialization", and that "the iterative enfolding of matter comes to matter" (Barad 2007: 180 et passim) does not establish clarity about what Barad is talking about and whether it may qualify as true.

VII

Before concluding, we need to look at one particular example that Barad discusses at great length toward the end of her book: the brittlestar. The example is truly important since it functions as a showpiece for Barad's posthumanism – the nearest she comes to offering a positive case for it.

To appreciate the role this example plays in Barad's agential realism, note that "phenomena", while always disclosed as an effect of particular intra-active engagements, "do not require cognizing minds for their existence; on the contrary, 'minds' are themselves material phenomena that emerge through specific intra-actions. Phenomena are real material beings" (Barad 2007: 361). Surprisingly, in speaking of the mind as emerging "through specific intra-actions", Barad refrains from making any connection with the contemporary debate over the emergence of mind between proponents and critics of panpsychism. Rather, what is important for her agential realism is its commitment to a "realist conception of scientific practices"; specifically, what "technoscientific practices" help make manifest is "an expression of the objective existence of particular material phenomena". Again, Barad's is not a realism committed to objectivity understood as referring to something in the world "preexisting" attempts to access it; rather, objectivity is "a matter of accountability for what materializes, for what comes to be"; and this, as we have seen, is determined by "which cuts are enacted" (Barad 2007: 361). To "enact the between" in Barad's intended sense is different from what Hacking is after in his notion of interference ("don't just peer, interfere"). "Intra-acting, not merely interfering", she writes, "is entailed in both experimental and theoretical practices" (Barad 2007: 359).

But what became of the issue of mind and matter as viewed in Barad's agential realism? First, consider what Barad takes the brittlestar to help illuminate:

"Brittlestars intra-act with their ocean environment and respond to differential stimuli made intelligible through these intra-actions, adjusting their positions and reworking their bodies in order to avoid predators or find food or shelter, all without brains or eyes" (Barad 2007: 379). What brittlestars help us realize, then, is that "the world is not merely an idea that exists in the human mind. To the contrary, 'mind' is a specific material configuration of the world, not necessarily coincident with a brain. Brain cells are not the only ones that hold memories, respond to stimuli, or think thoughts" (ibid.). Hence, Descartes' "I think, therefore I am" is "not the brittlestar's credo". Indeed, the brittlestar is a creature that demonstrates the flaws of the narrative about human exceptionalism. Barad uses it to advance a set of general claims: "Knowing is not a capacity that is the exclusive birthright of the human. The 'knower' cannot be assumed to be a self-contained rational human subject, nor even its prosthetically enhanced variant. There is no *res cogitans* that inhabits a given body with inherent boundaries differentiating self and other" (ibid.).

But what sort of creature is a brittlestar? And how well does it suit the agenda of Barad's agential realism?

The brittlestar caught Barad's attention when the *New York Times* ran the story "Eyeless Creature Turns Out to Be All Eyes" in 2001, summarizing the findings of a study published in the August 23, 2001, issue of *Nature*. An international team of multidisciplinary scientists had made the amazing discovery that the brainless and eyeless creature called the brittlestar, an invertebrate cousin of the starfish, sea urchin, and sea cucumber, has a secret: its entire skeleton forms a big eye. The species called *Ophiocoma wendtii* has "a skeleton that functions as a visual system, apparently furnishing the information that lets the animal see its surroundings and escape them" (Aizenberg et al., 2001; quoted in Barad 2007: 370). The Times journalist adds that "the brittlestar architecture is giving ideas to scientists who want to build tiny lenses for things like optical computing", prompting Michael Land, who studies animal vision at the University of Sussex in Brighton, England, to exclaim: "It's bizarre – there's nothing else that I know of that has lenses built into its general body surface" (ibid.). In interviews with the press, the study's lead author, Joanna Aizenberg, makes a "high-tech" comparison to a digital camera that builds up a picture pixel by pixel, explaining that in this marine creature "the calcic lenses surround the whole body, looking in all different directions and providing peripheral vision to the organism". This is what makes for its importance for optical devices, cameras in particular: "Instead of having one lens pointing in one direction, you could have thousands of lenses pointing in different directions. This will give you perhaps a 360-degree view of the whole space" (Aizenberg quoted in Barad 2007: 371).

What it is that makes the discovery of the brittlestar's capacity for vision so fascinating? It seems to be two things in particular: one is the frank astonishment expressed by scientists over what such a tiny marine creature is capable of; the other is the prospects that this discovery is hoped to hold for future scientific research that can be put to human technological and commercial use.

In the case of the brittlestar, then, we encounter an organic creature that can manipulate inorganic matter with truly impressive precision, doing so without having a brain. Considering that human ingenuity came up with microlens only a few years ago, and that photonic materials are to be found in opals and in the wings of butterflies, it begins to dawn on us – or at least on the scientists reflecting on the case – that nature, time and time again, foreshadowed mankind's technological developments, that "nature has been in the business of developing optical structures for a very long time", to quote, as does Barad, the physicist and optics specialist Roy Sambles (Barad 2007: 372). Yet Sambles' remark prompts Barad to comment that "the brittlestar may not get full credit for its superior ingenuity, which exceeds the current technological ingenuity of humans, but a larger, older, and wiser configuration called 'nature' does". She then cites a National Public Radio reporter: "Even the most primitive creatures might have the edge over modern science", leading Barad to quip "so what makes it 'primitive' again?" (Barad 2007: 373).

Barad does well to dwell on the upbeat responses that the "discovery" of the brittlestar's "secret" in terms of visual capacities has prompted in the scientific community, intimately linked as it nowadays is to the technological one. Joanna Aizenberg is a case in point when she proceeds immediately from the evidence now found of the remarkable ability of organisms such as the brittlestar to "optimize one material for several functions", to speaking about the "new ideas for the fabrication of 'smart' materials" that this discovery provides her team at Bell Laboratories (owned by Lucent Technologies) with. And indeed, telecommunications networks and computers now stand to benefit through "improved optical lithography techniques", bearing out the aspiration of scientists to "mimic nature's success and design microlenses based on the brittlestar model". As John Rogers, director of nanotechnology research at Bell Laboratories points out, "biomimetics builds on nature's expertise", highlighting that, as exemplified by the brittlestar, "a relatively simple organism has a solution to a very complex problem in optics and materials design" (Barad 2007: 374). What we can hope to learn by studying such biological organisms, then, is a technique that may be useful in nanotechnology. In fact, so great is the enthusiasm sparked by the study of the brittlestar that nothing less than a revolution in manufacturing optical devices is announced in its wake: "The new Bell Labs approach may revolutionize how crystals are made in the future for a wide variety of applications" (ibid.).

What does Barad make of the "discovery" of the brittlestar and the responses it has led to among the scholars involved? Again, what is the relevance of the brittlestar for her agential realism?

First, the brittlestar's optical system, correctly understood, undercuts the various premises on which the history of Western epistemology – championing representationalism and a clear-cut division between knower and known – relies. The brittlestar, Barad observes, "does not have a lens serving as the line of separation, the mediator between the mind of the knowing subject and the materiality of the outside world". The bottom line is this: "Brittlestars don't *have* eyes; they *are* eyes" (Barad 2007: 375). One is reminded of Merleau-Ponty's observing as much with respect to the human body. And yet, Barad maintains, it is not merely the case that the brittlestar's visual system is embodied; the deeper point is that "its very being *is* a visualizing apparatus" (ibid.). Since it is a "living, breathing, metamorphosing optical system", for a brittlestar, "being and knowing, materiality and intelligibility, substance and form, entail one another". And since we have to do with an animal without a brain, "there is no *res cogitans* agonizing about the postulated gap (of its own making) between itself and *res extensa*"; moreover, "no optics of mediation, no noumena-phenomena distinction, no question of representation" (Barad 2007: 375).

We begin to realize why Barad has selected the brittlestar as a showpiece for her agential realism: hand-in-glove, it illustrates a series of claims otherwise advanced in a very theoretical and general manner. The impression conveyed is that everything that can be said on behalf of agential realism can be said with respect to the brittlestar, and vice versa. In addition to the statements to this effect just referred to, there is the contention that the brittlestar's "*discursive practices* – the boundary-drawing practices by which it differentiates itself from the environment with which it intra-acts and by which it makes sense of its world, enabling it to discern a predator, for example – *are materially enacted*" (Barad 2007: 375; italics in original). Similarly, the brittlestar's "bodily materiality is not a passive, blank surface awaiting the imprint of culture or history to give it meaning or open it to change; its very substance is morphologically active and generative and plays an agentive role in its differential production, its ongoing materialization" (Barad 2007: 375f.).

It does not stop there, however. The brittlestar, Barad tells us, also helps challenge disembodied epistemologies as well as nontraditional notions of embodiment, demonstrating that "bodies are not situated *in* the world; they are part of the world" (ibid.). More spectacularly, brittlestar species exhibit tremendous diversity in sexual behavior and reproduction, leading Barad to ask such questions as: "When is a broken-off limb only a piece of the environment, and when is it an offspring? At what point does the 'disconnected' limb belong to the 'environment' rather than the 'brittlestar'?" (Barad 2007: 377). Clearly intrigued by all the questions and possibilities opened up by the breathtaking richness of brittlestar agency, Barad ups the ante, as it were, entertaining increasingly bolder ideas, such as: "Imagine the possibilities for lost limb trauma when it comes to brittlestars! Rethinking embodiment in this way will surely require rethinking psychoanalysis as well" (ibid.).

Now, while Barad is correct to point out that "brain cells are not the only ones that hold memories, respond to stimuli, or think thoughts" (Barad 2007: 379), suggesting that nonhuman creatures such as the brittlestar may do all of this as well as humans, psychoanalysis is a good candidate for a practice whose enacting as well as rethinking human agents alone seems to be eligible for. I am not sure about Barad, but what I for one find most fascinating and important about the comparison between brittlestars and humans that Barad invites us to examine, is certainly the differences that deserve to be recognized no less than the (more or less spectacular and sensationalist) similarities. Yes, like us humans – only more so, definitely better so – brittlestars can see. And yes, like us – only more so, arguably better so – they can fundamentally change, modify, and exploit their environment, to the point, even, when what that environment is – where it starts and where it ends – apart from the individual brittlestar engaged in effecting the changes, becomes impossible to tell: the capabilities of the brittlestar are such as to *physically dissolve* the conditions for answering the question in the clear-cut, division-drawing manner we take for granted will apply.

Ironically, in invoking psychoanalytic "lost limb memory trauma" to suggest that brittlestars may well be more like us than we are used to thinking, Barad contradicts her own plea about "attentiveness to differences that matter" (Barad

2007: 382). Clearly psychoanalysis is a uniquely human invention, concept, and practice if there ever was one. Equally clearly, there are features and capabilities in brittlestars that have no match in humans and that show a superiority on their (the nonhuman organism's) part. Witness, to take my favorite example, how good birds are at flying as compared to humans – on this score (and what could be more important?) humans are no doubt a sorry and slow species judged from the bird's literally higher point of view. Barad is therefore spot-on when she criticizes the presumably "democratizing" move of "inviting nonhuman entities into our sociality", arguing – correctly – that "the culture-nature dualism is not undermined by inviting everything into one category (man's, yet again)" (Barad 2007: 378).

Brittlestars, Barad rightly emphasizes, are not "pure bits of nature or blank slates for the imprinting of culture"; nor are they "mere resources or tools for human interventions". Rather, they are agentive beings, "with more entanglements than arms". "Humans" and "brittlestars", Barad continues (the scare quotes are hers), "learn about and co-constitute each other through a variety of brittlestar-human interactions". Now biomimesis, as carried out by the scientists quoted above, may be a research program that seeks to appropriate the ingenuity of the brittlestar's lens system, "but this practice cannot be understood as a process of copying the other". Why not? Because "nature is not a pure essence that exists 'out there' or on a slide positioned under the objectives of our microscopes" (Barad 2007: 382).

Despite the validity of Barad's last claim here, can one really say that humans and brittlestars "learn about and co-constitute each other" through so many interactions? This is a tricky question, and it is so for several reasons, some ontological, others ethical. Barad is explicitly accentuating the ethical dimension in the last section of her book where the brittlestar case is discussed.

So how to answer the question? In suggesting a two-way relationship, one of co-constitution, between humans and brittlestars, stressing that processes of learning will go both ways and so presumably benefit both parties, Barad surely wants to avoid the anthropocentric fallacy of portraying it as a one-sided affair, where humans are the active part and the nonhuman creature the passive one: the subject-object, culture-nature dualisms are duly laid to rest. Yet I find this posthumanist stance problematic. The ethically motivated (and as such unobjectionable) emphasis on co-constitution and interaction, premised on the notion that both parties have something to learn from each other, suggests a kind of *reciprocity and symmetry* that is simply misleading as to how the current humans-brittlestars regime in fact is, like it or not. The ethical "ought" comes to the fore and dictates the description offered and the words employed in a way that risks distorting the factual "is", pervaded as it (however regrettably) is by the gross asymmetry of power between the parties. True, Barad acknowledges that "these echinoderms don't reflect on the world" (Barad 2007: 382). But she fails to address the questions raised by this tremendously important difference in being-in-the-world between humans and brittlestars.

Without being an expert on brittlestars, and without knowing much about the research now enthusiastically carried out under the heading of "biomimesis", what I am struck by is that the interest humans (researchers, technologists, investors, even politicians) take in brittlestars is a very one-sided affair. We have not asked the brittlestars about how this interest of ours in them, based on our discovery of their "secret", i.e., their capacities for vision, are perceived from their brittlestar point of view. What if their response goes like this: "Hey, researcher and investor, leave us brittlestars alone!" Pondering their response may be a difficult thing to do, of course; after all, we are only humans, so we certainly have our limits, right? Yet, while there are obvious limits to our chances of understanding brittlestars and seeing the world, including what we are in the process of doing to them, from their point of view, it is not for that (epistemic) reason something that does not count ethically. On the contrary: due to the lack of possibility for actual reciprocity in communication and exchange between the two parties involved here, we humans are carrying an even larger burden of argument for what we are doing to/with these creatures than we would in a case where the affected party were capable of responding to our doings in a way that we could easily comprehend, and where our unwillingness to do so in practice would backfire on us. Asymmetry in power and capabilities is the very core of ethical responsibility: the more clear and undeniable it is that you are the strongest party, the greater the responsibility.

Since, as we shall see in a moment, Barad explicitly commits herself to the Levinas-inspired ethics of responsibility that I have just alluded to, her discussion of biomimesis is illuminating in a way that is both surprising and disturbing. She starts by saying that "it would be a serious error to mistake biomimesis for mere imitation". Far from being about "copies of originals", or even "copies of copies", the merging field of biomimesis is to be looked upon as "a particularly poignant call for the incorporation of difference at every level in breaking the deadening and sinister symmetry of Sameness that uses the hall of mirrors to suck time, history, and matter into the black hole of stasis (leaving in its stead a culture of no culture and a nature of no nature)" (Barad 2007: 382; the notions in the last formulation are borrowed from Donna Haraway 1997). For all the force of this rhetoric about "Sameness", I cannot help but think that Barad is barking up the wrong tree. She assures us contemporary practitioners of biomimesis "do not claim to be making replicas of nature; rather, they are engaged in practices that use nature as inspiration for new engineering designs". This being so, biomimetics "honors Mother Nature as the primo engineer, but it doesn't promise to abide by her methods. It embraces new innovations, new materials, new techniques, new applications. Bringing the new to light is its highest principle" (ibid.).

This may well be a correct portrait of what the new "bio-info-nana-technologies" are up to as they "embrace the new for very practical reasons", new engineering designs being one of them. It would, however, be an equally valid portrait of what contemporary capitalism is up to as we are entering the Anthropocene. Indeed, there is story to tell about the importance of the fact that as of today "30 percent of the United States' gross national product is said to derive from technologies based on quantum theory" (including cell phones and portable computers) (Barad 2007: 252). But Barad is content to mention this fact only in passing; she does not use the technological, entrepreneurial, and commercial interest presently taken in the brittlestars as an opportunity to dwell on it, making her account of these developments surprisingly uncritical. Thus, Barad observes, matter-of-factly and without comment, that "quantum computing represents an important possibility for maintaining our competitive edge" (Barad 2007: 386); moreover, she notes the "current fascination with biomimetics" and the "enthusiastic support it is receiving from government agencies, universities, and private industry", anticipates the opening up of "possibilities for forming partnerships with brittlestars and other organisms for biomimetic projects" and considers all of this to illustrate how "the entanglements we are part of reconfigure our beings, our psyches, our imaginings, our institutions, our societies; 'we' are an inextricable part of what gets reworked in our R & D projects" (Barad 2007: 383).

Yet the "we" invoked here, for whom the potential for "reconfiguring" will hopefully prove beneficial and worthy of all the enthusiasm and investment mentioned, strikes me as a remarkably one-sided and amputated one, echoing the anthropocentrism betrayed in calling the brittlestar's capacities of vision its "secret" – surely a description of a nonhuman creature that flows directly from the human observer's objectifying point of view, suggesting that if it weren't for the way in which that creature's vision qualifies as comparatively superior to that of us humans, the fascination, not to mention research-based investments, would not have existed – nor, presumably, would the brittlestar have figured in Barad's book.

Given this background, it is amazing that Barad considers what she describes as "our intra-actions with brittlestars" to be a suitable case for highlighting what ethics is all about: Ethics, she contends, "is about mattering, about taking account of the entangled materializations of which we are a part, including new configurations, new subjectivities, new possibilities – even the smallest cuts matter". Hence we are "much more intimately connected than the notion of mimesis connotes" since "we don't have the distances of space, time, and matter required to replicate 'what is'" (Barad 2007: 384). So what follows, ethically speaking? Barad's answer is that "the ethical practice of biomimesis will require specific case-by-case accountings for marks on bodies. Technoscientific practices are about making different worldly entanglements, and ethics is about accounting for our part of the entangled webs we weave" (ibid.).

These are very general statements, easy to sympathize with in principle, yet hardly helpful when it comes to deciding, say, how to deal with brittlestars so as to ensure that they are not treated simply as a means to human ends. In her book's last section, Barad turns to the philosopher Emmanuel Levinas to sketch "an ethics of mattering". What she finds attractive in Levinas's ethics is its emphasis on "the otherness of the Other" as given in responsibility, whereby responsibility is not understood conventionally as a commitment that a subject chooses freely but instead as "an incarnate relation that precedes the intentionality of consciousness". Barad takes Levinas's message to be that "we cannot ignore the full set of possibilities of alterity [. . .] including the 'other than human' as well as the 'human'", leading her to conclude that "responsibility – the ability to respond to the other – cannot be restricted to human-human encounters when the very boundaries and constitution of the 'human' are continually being reconfigured and 'our' role in these and other reconfigurings is precisely what 'we' have to face". Therefore, a "humanist ethics won't suffice when the 'face' of the other that is 'looking' back at me is all eyes, or has no eyes, or is otherwise unrecognizable in human terms. What is needed is a posthumanist ethics, an ethics of worlding" (Barad 2007: 392).

When I interviewed Levinas in 1993, however, there was no hint of support for such posthumanist ethics in his clarification of what it means to respond ethically – i.e., responsibly, without presupposing or demanding reciprocity – to what he famously refers to as "the face of the Other" (see Jodalen and

Vetlesen 1997: 53ff.). When Levinas asserts that "the relation to the face is straightaway ethical", so that "the face is what one cannot kill, or at least it is that whose *meaning* consists in saying: thou shalt not kill" (Levinas 1990: 89), he explicitly says that the relationship he is talking about is the "inter-human" one (1990: 97). That also holds for the scenario he explored in his *Otherwise than Being or Beyond Essence*, where, as soon as the so-called "third" appears on the scene, the question becomes: who is the Other and who is the third?, meaning: whom am I to prioritize given that I cannot commit myself equally to both? I have to make a choice, but based on what, to ensure it is ethical? For all the complexities Levinas recognized and struggled with in the case of the ethical triad as distinct from the dyad, what he took for granted and never questioned was that the subject as well as the Other (turning second) and the third were all humans.

There is one text by Levinas that is particularly relevant for Barad's proposed posthumanist reading of his ethics, namely his essay "Name of a Dog" (not mentioned by Barad). In it, Levinas tells the story of Bobby, a dog that Levinas got to know as a member of a unit of Jewish prisoners of war in Nazi Germany. Bobby's affectionate enthusiasm for the prisoners, his excited barking when they returned from work in the evening, prompted Levinas to call him "the last Kantian in Nazi Germany". Despite his affection for Bobby, however, Levinas argues that dogs are without ethics and without logos, thus reasserting, as Deborah Bird Rose points out, "an absolute boundary" between himself and Bobby, underscored by the claim that Bobby, being a dog, was "without the brain needed to universalize maxims and drives" (Levinas 1990: 153; Rose 2011: 30). In other words, for Levinas, Bobby lacks a "face", a lack that proves fatal within Levinasian ethics in that it spells the dog's (and presumably all other nonhuman animals's as well) exclusion from the moral universe of responsibility and thus from the sort of relationship where the command "thou shalt not kill!" obtains. I concur with Rose's comment that "Levinas's rejection of Bobby is anguishing because the greatest twentieth-century philosopher of ethical alterity could not unambiguously make a place in his ethics for the one living being who approached him and others with full recognition of their humanity" (Rose 2011: 30). Indeed, by invoking the distinctly Kantian ability to universalize maxims, Levinas "restores the tyranny of the abstract over the living reality of the world". By taking logos to be that which fundamentally differentiates humans from animals, Levinas not only echoes Kant's intellectualism; he also "reinscribes the big dualisms of Western thought – mind over and above matter, the abstract and eternal over the living reality in all its immediacy, proximity, and dynamic transience" (Rose 2011: 31). Based on his rejection of Bobby, then, Levinas would counter rather than endorse Barad's statement that "just as the human subject is not the locus of knowing, neither is it the locus of ethicality" (Barad 2007: 393).

There is more. When Barad maintains that "ethics is not about right response to a radically exterior/ized other, but about responsibility and accountability for the lively relationalities of becoming of which we are a part" (ibid.), I see three problems. The first is that Barad's emphasis on how everything is connected and entangled with each other (be it brittlestars and humans) glosses over exteriority and otherness in a way that is especially unfortunate from an ethical point of view. In fact the strong point in Levinas's ethics – bracketing for now its anthropocentricity – is its insistence that the Other is, and is allowed to remain, *different*: not only allowed to *be* different, but to *flourish in that difference* from the ethical agent who – preintentionally and ineluctably, without having asked for it, or even thought about it – is responsible when confronted with the otherness of the Other. As readers of Levinas will know, it is precisely to underscore the ethical primacy of otherness (also designated as "exteriority" and "alterity") that he uses the capital O to refer to the Other.

The second problem has to do with Barad's stated agential realist aim of "rethinking and reworking causality in terms of intra-activity", asserting that "events and things do not occupy particular positions in space and time; rather, space, time, and matter are iteratively produced and performed". The ethical importance of this is that "consequentiality, responsibility, and accountability take on entirely new valences. There are no singular causes. And there are no individual agents of change. Responsibility is not ours alone. And yet our responsibility is greater than it would be if it were ours alone" (Barad 2007: 394). Leave aside for now the way these statements risk concealing and ignoring asymmetries of power, such as obtain – routinely, structurally, and institutionally, in contemporary Western society – between humans and nonhumans (such as brittlestars). The "all-included, nobody left out" rhetoric of intra-activity and intra-entanglements, combined with the downplaying of individual agency, may seem well suited to articulate the holistic stance that "everything is connected to everything", even brittlestars to humans. But as Deborah Bird Rose and her fellow animals studies scholar Thom van Door en point out, everything is connected to *something* (something particular, at some particular time and place), which is connected to something else, in the same particular – concretely situated – manner, so that "while we may all *ultimately* be connected to one another, the specificity and proximity of connections matter – who we are bound up with and in what ways" (van Dooren 2014: 60).

Finally, the third problem concerns the practical consequences of the above objections. Barad, it will be recalled, warns against what she refers to as the "move to invite nonhuman entities into our sociality" (Barad 2007: 378), presumably because it would amount to an erasure of differences (say, those between brittlestars and humans). Yet differences live dangerously within Barad's agential realist universe. Philosophically, this is to a large extent due to Barad's critical, posthumanist stance toward the series of dualisms – mind versus matter, humans versus animals, the organic versus the inorganic, etc. – that has pervaded Western thought and science; dualisms that are not only lofty philosophical such, but that have been mobilized to justify various practices of discrimination, abuse, and downright eradication of the "inferior" or "unworthy" entities on the wrong side of the division. I am not suggesting that the critique that these dualisms and their practical abuse have received is ill-founded or mistaken – much of it is thoroughly justified. But differences must be acknowledged; distinctions must be made – we must avoid throwing the baby out with the bath water just because they have been exploited for ideological reasons to help justify deeply unethical practices. Such practices, however, are enacted by and wedded to particular agents, making decisions and performing actions that affect particular others, be they human or nonhuman. Otherwise, responsibility, the ethical category *par excellence*, fails to have purchase in the particular case – and there are no others.

4 Animism – panpsychism in practice

In this chapter, I turn to major contributions in recent anthropology to develop my thesis that animism be regarded as panpsychism in practice. My thesis does not imply a simple one-to-one relationship between what the philosophers we have discussed theorize as panpsychism and what the anthropologists doing fieldwork among indigenous peoples depict as so many particular ways of practicing an animistic cosmology, one where everything that is real is treated as an embodiment of agency, indeed, as so many persons in so vast a manifold of different guises. The upshot is that the conventional contrast between humans and nonhumans – considered pivotal in so many respects – is being challenged, this time not from philosophers rocking the boat of the Western canon but from anthropologists whose experiences among people enacting an alternative cosmology to the anthropocentric one that I consider to be of the highest relevance, indeed urgency, now that the very constellation "nonmodern" peoples/cosmologies is threatened with extinction.

I

In 1990, the French philosopher of science Michel Serres published a small monograph entitled *The Natural Contract*. At first barely noticed outside France, the book has slowly gained reputation as a classic and is often referred to by prominent scholars such as Bruno Latour. It contains a passage that may have appeared speculative at the time of writing, yet will strike today's readers as stating a fundamental fact:

> As of today, the Earth is quaking anew: not because it shifts and moves in its restless, wise orbit, not because it is changing, from its deep plates to its

> envelope of air, but because it is being transformed by our doing. Nature acted as a reference point for ancient law and for modern science because it had no subject; objectivity in the legal sense, as in the scientific sense, emanated from a space without man, which did not depend on us and on which we depended de jure and de facto. Yet henceforth it depends so much on us that it is shaking and that we too are worried by this deviation from expected equilibria. We are disturbing the Earth and making it quake! Now it has a subject once again.
>
> (Serres 1995: 86)

Something unprecedented is taking place: "the Earth is moved". Hence "the immemorial, fixed Earth, which provided the conditions and foundations of our lives, is moving, the fundamental Earth is trembling" (ibid.).

What is happening? And why? It is a large claim to say that the Earth is "being transformed by our doing". What kind of doing does Serres have in mind? What kind of agents and what kind of actions?

Serres may well be the wrong author to ask, better at sounding the alarm bell than explaining how it came about that "we" are presently disturbing the Earth. He observes that "when our tools were local, and kept us working on only our own hayfield, we weren't constantly informed of the Earth's global changes" (1995: 109). True enough. But the point he raises to such dramatic effect in the cited passage is that in a preindustrial, low-tech era when the world perceived and experienced was local not global, there were no "global changes" affecting the Earth to talk about – the changes brought by "our" disturbance had not kicked in yet.

To be sure, Serres does point to the role played by "our technologies", helping bring about today's situation where "the global power of our new tools is giving us the Earth as a partner, one whom we ceaselessly inform with our movements and energies, and who, in return, informs us of its global change by the same means" (1995: 109). What does this partnership consist in? "Equipotent to the Earth", writes Serres, "we have become its biplanet, and it is likewise becoming our biplanet, both bound up by an entire planet of relations. A new revolution, in the Copernican sense, for our grandeur and our responsibilities. The natural contract resembles a marriage contract, for worse and for better" (1995: 110).

So, to allude to Heidegger's famous take on Hölderlin: "Where danger is, grows the saving power also" ("Wo aber Gefahr ist, wächst das Rettende auch") (Heidegger 1977: 42): a new relationship between us and the Earth can be built. Serres strikes a hopeful note: "Here, then, is nature today, new and fresh, being born: global, whole, and historiated before the eyes of global humanity as a whole"; "concrete and technological right now, since our means of intervention act on it and it in turn acts on us; a network of multiple bonds where all things, congruent, conspire and consent; a web tied, by a lattice of relations, to the henceforth united social and human fabric" (1995: 111). Thus, while it is "only in this century that nature has been born, really, before our eyes, at the same time as a humanity bound in real solidarity" (ibid.). Whereas what Serres calls "the modern social contract" is, or was until now, "unaware of nature", meaning that nature was simply taken for granted, as plentiful, harmonious, and constant at that, offering up what any specific human society, locally and limitedly, needed to take from it, the "natural contract" that Serres now contends sees the light of day is "metaphysical" because it "goes beyond the ordinary limitations of the various local specialities". What is truly important about this emerging natural contract, however, is that it "acknowledges above all the new equality between the force of our global interventions and the globality of the world" (1995: 46).

If this strikes a note of optimism, a few pages later the tenor is decidedly less upbeat. The "new equality" that Serres just spoke about may turn out to be more of a normative ideal than the currently emerging historical reality: the respect for nature as an equal partner in a supposedly symmetric relationship with a global humanity "bound in real solidarity" does not materialize. To the contrary, "modernity neglects, speaking in absolute terms. It cannot and will not think or act toward the global, whether temporal or spatial" (1995: 48).

This is French philosophy, and whatever its merits, the claims made by Serres are on a very high altitude of abstraction and generalization: modernity is a subject, an agent, unified and homogeneous, that neglects, that cannot and will not act in certain ways. This subject fails to do what Serres' notion of the natural contract considers necessary to establish a good marriage between humanity and nature, to use Serres' analogy. How has this failure come about? This is what Serres tells us:

> Through exclusively social contracts, we have abandoned the bond that connects us to the world, the one that binds the time passing and flowing to the weather outside, the bond that relates the social sciences to the sciences of the universe, history to geography, law to nature, politics to physics, the bond that allows our language to communicate with mute, passive, obscure

> things – things that, because of our excesses, are recovering voice, presence, activity, light. We can no longer neglect this bond.
>
> (1995: 48)

So there has been a kind of fall, historically, in that the bond that once connected us to the world, to nature, has been lost. It is not clear why that happened, what sort of forces – societal, economic, cultural, political – that helped cause it, but we already saw that technology – the shift from local to global, the expansion in time and space fueled by ever-growing efficiency in methods and means – plays a central part in Serres' narrative. More important, however, is the ambiguity of the statement addressing "mute, passive, obscure things". What are they? Are they *made* to be that way, having once been otherwise: vocal, thriving, active entities in the world? Is Serres referring to creatures with agency in that sense, only that its manifestation has been thwarted, denied them? And in talking about "obscure" things – obscure in what sense, for whom, and as perceived from whose point of view? Is this a Serresian equivalent to Barad's brittlestar, whose visual capacities she described as its "secret" – secret for whom?

Serres does not tell us, so we simply do not know if the things he has in mind are mute, passive, and obscure in-themselves, or for us; *en-soi* or *pour-soi*, to invoke Griffin's use of Sartre's distinction. More to the point, we do not know whether they have been subject to substantial change in their very being, affecting their distinct properties, as a consequence of the "disturbance" to nature that Serres speaks about in a most general manner. Even more unclear, though, is Serres' assertion that the things he refers to, "because of our excesses, are recovering voice, presence, activity, light", taking this to help us realize that "we can no longer neglect this bond".

I find this suggestion counterintuitive. If anything, do "our excesses" not contribute to the silencing, or downright extinction, of the creatures whose voice Serres talks about, rather than its recovery? Or is he playing, again, with the dialectic of danger and chance, so that the moment these creatures are endangered will turn out to be their chance to regain their voice, if only we allow them to? Yet he starts his book by painting a picture where such an option is all but ruled out. Serres states flatly that "once victorious, the Earth is now a victim"; "conquered, the world is finally conquering us" (1995: 11, 12). And whereas the Earth could well exist without us, "we cannot exist without it". What follows? That we "must indeed place things at the center and us at the periphery, or better still, things all around and us within them like parasites" (1995: 33). Echoing Horkheimer and Adorno's classic work of sweeping culture critique, *Dialectic of Enlightenment*, Serres warns that "through our mastery, we have become so much and so little masters of the Earth that it once again threatens to master us in turn". Indeed, pure mastery – "unregulated, exceeding its purpose" – is turning back on itself, requiring us to "master our mastery" (1995: 34). Our culture has for too long "abhorred the world"; earth, water, and climate, "all those things that never interested anyone, from now on thrust themselves brutally and without warning into our schemes and maneuvers. They burst into our culture, which had never formed anything but a local, vague, and cosmetic idea of them: nature" (1995: 3).

Eloquence and rhetorical force notwithstanding, there is something utterly provincial about the viewpoint from which Serres advances his series of generalizations about the relationship between humans and nature. To suggest that earth, water, and climate are "things that never interested anyone" is a contestable way of writing history, to put it mildly. To the contrary: the elements have provided the framework for human existence on planet Earth since our species emerged here; earth and water and the geographically specific seasonal rhythms in temperatures, in darkness and light, and so in determining access to the riches of the earth that humanity always and everywhere has been dependent on, have, on account of that very dependence – ours on the elements, not the elements on us – commanded entire cosmologies of an animist, panvitalist, or panpsychic kind, instilling in humans the sort of awe and wonder, combined with humility and fear, that modern thinkers like Descartes found so incompatible with scientific inquiry as to debunk it as an entirely "irrational" stance toward "occult" forces. As "moderns", and in the name of progress, science, and rationality, we have so effectively internalized the mechanistic model of nature inaugurated by Newton, Bacon, Descartes, and Hobbes that we forget how radical a break it represents with the worldview that preceded it and that historically is the rule not the exception – mechanism is – as Hans Jonas reminds us, observing that "*when man first began to interpret the nature of things, life was to him everywhere, and being the same as being alive*" (Jonas 2001: 7; my italics). *Animism* was the most widespread cosmology: "Soul flooded the whole of existence and encountered itself in all things". That the whole world is alive, Jonas notes, "is really the most natural view, and largely supported by prima-facie evidence" (ibid.). The inanimate "dead" matter of the Cartesian *res extensa* was yet to be discovered. To members of animistic cultures, the notion that nature is dead, that it is devoid of life and spirit, purposefulness and value, would be but completely unintelligible, the proposition of a madman. Panpsychism – comprising not only animals but earth, wind, and water, all the elements making up matter in the Cartesian sense – was experientially sustained by the actual preponderance of life in the horizon of its earthly home. Resonating deeply with the way in which nature was experienced, the omnipresence of life was considered an indubitable truth.

Only since the Renaissance has all existence come to be considered, treated, and acted upon as pure matter, stripped of all features of life; a field of inanimate masses and forces operating according to the laws of inertia and of quantitative distribution in space; laws that, according to Hobbes in particular, comprised not only the natural world but man's social and political one as well. Absolutely crucial to the hegemony of the "new science" was the banning of anthropomorphism (first encountered in the discussion of Nagel in Chapter 1), making sure to remove all manifestations of man's inclination to project (as the new psychology would have it) his own felt aliveness, his state of mind or feeling, onto the physical and as such indifferent – "neutral"- physical reality. Only properties allowing for measurement, for comparison, calculation, prediction, and control within a wholly quantifiable framework qualified as apt for the requirements of the "exact" knowledge that became attainable once the allegedly primitive and superstitious notion of "occult" forces (spirits, manifestations of mentality, of intelligence, of purposefulness) had been abandoned – abandoned not only from the carrying out of science but from everyday life and culture in general. Jonas credits Kepler with being the first among the moderns to declare quantity the one essential and truly knowable aspect of reality; henceforth, cognition consisted in measurement and comparison of measurements, ideally carried out in the controlled setting of the laboratory experiment which allowed, in principle, for the repeatability of results, the reproducibility of observations, and the exchangeability (anonymity) of the observers – the very notion of how experiments must be carried out in order to qualify as scientific that we saw Bohr persist in subscribing to, notwithstanding the revolution in modern physics that he promoted. In this way, *method gained primacy over subject matter*, epistemology over ontology, to echo Charles Taylor (1985, 1995): the formal-mathematical, abstract and universally applicable requirement of measurability came to define, indeed to exhaust, the domain of the knowable, banning the qualitative aspects of life in its manifoldness and diversity so completely as to render *the lifeless* definitional of what could be strictly known, and so tantamount to reality as such (see Jonas 2001; Merchant 1980, 2016).

II

What led me to make these observations was Serres' claim that "earth, water, and climate" are "things that never interested anyone" (1995: 3). The claim ignores the fact that the dominant cosmology in human history, with the exception of the last three or four centuries, has been variations of animism. As the French anthropologist Philippe Descola shows in his magisterial *Beyond Nature and Culture*, the modern cultural and scientific dualism according to which humans and nonhumans exist in separate domains marks "the absolutely exceptional nature of modern cosmology" (2013: 87). The essential difference, understood as an ontological abyss, between humans and the nonhuman world of plants, animals, geology, and natural forces is not only a specifically Western notion but also a very recent one. On this dualism, "natural" objects have no interiority: no mental and psychic properties, no intentionality, no ability to adopt an agential point of view in/on the world, no way of having things – what comes to pass – *matter to them* and responding in kind by way of seeking to optimize and secure what is perceived as desirable and life-sustaining, and no way of avoiding or overcoming what is perceived as the opposite. Whereas, as Descola shows, a modern naturalist cosmology denies such properties and the faculties and skills that go with them to nonhumans, animist cosmologies admit them to all creatures – not only animals – insofar as they are manifestations of life and hence of the fundamental striving to persist (*conatus*). Hence, animism and naturalism are strictly antithetical ways of discerning the properties of things: animism highlights the physical differences between things; it emphasizes that they have dissimilar bodies but insists, nevertheless, that they maintain similar interrelations in that they – humans and nonhumans alike – share an analogous interiority, their differences in interiority – i.e., only humans adhere to moral norms, make promises, can ask each other for forgiveness, etc. – being seen along a continuum, as gradual rather than absolute, and so maintaining that all are agents and must be treated as such. In stark contrast, naturalism highlights the physical continuity between the world's elements that follows from all being subject to the laws of nature, insisting, nevertheless, on the heterogeneity of the relations that may bring them together, depending on their capacity or lack of such to manifest interiorities of various kinds. The differences in physical dispositions between humans and nonhumans that animism pays such attention to, does not in any way prevent its practitioners – such as an (Upper Amazon) Achuar hunter – from engaging in "a permanent dialogue of souls, and this intersubjective communication is the basis for the principle of an unrestricted sociability that encompasses both humans and nonhumans in its universal network" (393).

Naturalism, by contrast, instead of acknowledging all creatures, no matter their differences of physical dispositions (pertaining in particular to styles and powers of movement), as participants in all-encompassing sociability, pushes all nonhumans to "the peripheries of collectives on account of their lack of any reflective consciousness and moral sense" (396), these being the prerogatives of humans that lend them superiority in but all relevant respects: cognitively, emotionally, volitionally, morally, legally – you name it. Being objects, not subjects, one does not enter into contact with plants or animals. And should anyone do so – perceive and treat nonhumans as persons on an equal standing with oneself qua human – it would not be regarded as an innocent mistake, a neutral error, for which one may be easily excused. This holds for adults; a child might be excused, seen as "not knowing better", that is to say, as not yet having internalized his or her culture's way of making the all-important distinction between the two entities in question. Should this prove a consistent pattern, not a one-off incident, such an adult person in a present-day Western society would be seen as manifesting deviant behavior, as making a serious category mistake in treating as similar (one's equals) what are essentially different (one's unequals), thus prompting the need for a well-considered response on behalf of the culture/cosmology contradicted – typically, requesting psychiatry to deal with the person, trying to find out how the deviance came about and what can be done to treat it.

To return to Serres, I would not expect him to disagree with Descola's way of distinguishing between animism and naturalism, between premodern (or better, nonmodern) and modern cosmologies. Nor would he, presumably, contradict my point about how present-day Western culture tends to individualize and pathologize what might well be deliberate acts of, say, greeting and communicating with a tree, a creek, or a wild animal encountered in the woods. Deliberate in what sense, exactly? In the sense of greeting the "natural object" as a person. Now, a proponent of the hegemonic culture intent on showing the inappropriateness of such behavior will typically ask: "And what was the response you got?" He will take the question to be unanswerable and hence to identify what is simply, objectively, a mistake on the part of the tree-greeter. Yet this is the very moment he reveals himself to have missed the point: the point is that nonhuman others have *other ways* of initiating contact as well as responding, of taking part in communication, than do we humans when dealing with each other. The counter-question is: Why on earth – precisely: on this one earth – should *that* way of communicating be presupposed as the only one possible, the only one rational and appropriate to the other at hand, and so exhausting the whole domain of meaningful and as it were non-pathological encounters and practices of greeting? Cross-species difference *is* the point. *Otherness* goes both ways: our from the tree, the creek, the wolf; and, respectively, their from us. I take Descola to allude to this when he writes that "the thoughtless ransacking of the planet's resources and the destruction of its biotic diversity [. . .] result from our forgetting the belief that prevailed in the first ages of modernity, namely that the splendid otherness of nature is necessary for the manifestation of the specific qualities of humanity" (397). This anticipates a point I shall return to below: in impoverishing otherness as found in our defined "others", we humans impoverish ourselves.

Again, it is a matter of not throwing the baby out with the bath water: on the one hand, to be willing to question the universality of the distribution of humans and of nonhumans into separate essential systems, a universality that is not borne out by the findings of anthropological research, being instead a myth upheld by Western thought's unenlightened propensity to take its naturalist stance as the model cosmology, as *the* rational stance; and on the other hand, to recognize that even though humans and nonhumans do not fit neatly into two separate ontological, mutually exclusive, and quasi-autonomous domains, it does not follow that they are not genuinely different kinds of creatures, exhibiting very different capacities and powers of agency, making them highly unequal partners in whatever inter-species relationships and communications are taking place, inequalities in power and voice being justified by the ways in which a given culture exploits *factual* differences between humans and nonhumans by rendering them *hierarchical*, normatively as well as purely descriptively, so that what makes humans differ from "the rest", and especially in the West, is taken as evidence of their superiority: we pronounce ourselves as "better" than other species, other creatures, in that we have something they lack, do something they can't. In doing so, we are not at all the "impartial" judge weighing the significance of differences that we pride ourselves on being; on the contrary, the whole – cultural, intellectual, moral – exercise is tendentious and self-serving, as is revealed by the way the undeniable capacities-based superiority of animals – birds can fly, humans can't – are ruled out as relevant to the distribution of positions in the hierarchies of power, status, and rights.

To appreciate how profoundly Western moderns' perception of "natural objects" differs from how nonhumans, in all their heterogeneity, have been perceived in nonmodern societies (Descola's preferred term), a closer look at animism is called for.

As indicated, animist systems use manifest differences in physicalities to introduce discontinuity into a universe peopled by persons with very disparate outward appearances (some have wings; some have eight, six, four or two legs; some have fins, some that are miniscule and some that are enormous). Yet at the same time, animist systems cast all of them as "so human in their motivations, feelings, and behavior" (Descola 2013: 130). What is the nature, and the overall significance, of this difference? According to Descola, it consists in the *form* and the mode of life that it prompts, and not in substance, as we are culturally socialized into thinking. Common to most animist ontologies, then, is the idea of a material continuity linking all organisms together. The corporeal form is therefore what differentiates between humans and nonhumans, since "the soul that all of them possess could not perform that function" (131). As the anthropologist Signe Howell observes about the Chewong of Malaysia, "consciousness in this sense makes one a 'personage' . . . regardless of one's outer shape (or 'cloak', in Chewong parlance), be it that of a gibbon, human, wild pig, frog, ramboutan fruit, bamboo leaf, the thunder-being, a specific boulder or whatever" (Howell quoted in Descola 2013: 131). Descola points out that the question of the discontinuity of bodies is the "obsessive" theme that Amerindian myths convey at every opportunity. There will be numerous stories about a time when humans and nonhumans were not differentiated. In those days, "animals and plants were masters of all the skills of civilization, communicated with one another with no difficulty, and abided by the major principles of social etiquette. As far as one can tell, their appearance was human, and only a few clues, such as their names and their strange behaviour, indicated what they were to change into" (131f.).

The overall point is that Amerindian myths do not evoke an irreversible switch from nature to culture. Rather, Descola explains

> they portray the emergence of "natural" discontinuities from an original "cultural" continuum within which humans and nonhumans were not clearly distinguished. However, this great movement of speciation does not result in a natural order identical to the one familiar to us, since, even if plants and animals now possess physicalities different from those of humans [. . .] most

> of them have so far preserved the faculties that they enjoyed before they split into different species. These faculties were subjectivity, reflective consciousness, intentionality, the ability to communicate in a universal language, and so on. They are thus persons, clothed in the body of an animal or a plant, which they occasionally set aside in order to live a collective life analogous to that of humans.
>
> (Descola 2013: 132)

The Amerindian peoples that Descola has in mind are well aware that, say, a jaguar or a peccary are indeed different from humans – and from each other – in the way that they orient themselves in the world. So in what sense are they persons – or, to put it more accurately, is their way of behaving that would cast them as persons? Descola's answer is intriguing: It is the jaguar and the peccary themselves, these Amerindian peoples hold, "who see themselves as performing the very same gestures as humans and who, in all good faith, fancy that they share with the latter the same technology, the same social existence, and the same aspirations" (132). The upshot is that, in their myths as well as in their daily lives, Amerindians "do not regard what we call culture as the prerogative solely of humans, since there are many animals and even plants that are reputed to believe that they possess a culture and live in accordance with its norms" (132).

Viewing the "natural order as a moral order", an all-inclusive one at that, is a commonplace among nonmodern cosmologies (see Sahlins 1976: 113). Yet the notion that animals and plants "possess a culture and live in accordance with its norms" appears to be but totally foreign to a philosopher like Serres: for all his critique of the anthropocentrism pervading what he calls "our culture" and the social contract it advocates, to the detriment of nonhumans of all stripes, Serres stops short of questioning the typically modern idea that *culture is the prerogative solely of humans*. It may be tempting to reply, on Serres' behalf, that even if it is granted that many animals and plants believe that they possess a culture (to paraphrase Descola), they are simply wrong in believing so. Yet pointing this out would not qualify as an objection, since if such other-than-human beings can be wrong in this manner, that very possibility presupposes the possession of mental capacities – for believing something to be the case, and acting upon that belief – that the claim that "culture is a prerogative of humans only" denies them.

That said, a more serious objection may be launched: bracket any notion about what animals and plants believe or not; concentrate instead on the question whether the Amerindian peoples – cultures, as it were – are justified in believing that animals believe that they possess a culture in said sense. This (the objection continues) would shift the burden of argument to where it belongs, namely to the wholly and exclusively human domain and so would be safely within the framework of anthropocentrism rather than contradict it. Talk about beliefs, then, is per definition talk about something that only humans are engaged in; only with regard to humans does the distinction between true and false beliefs make sense.

If this sounds familiar and helps us return to dry land, and it certainly does, then that is proof that something got lost when we turned from the anthropologist Descola and back to the philosopher Serres. What got lost?

III

When the Brazilian anthropologist Eduardo Viveiros de Castro studied the Amerindian ethnonyms usually translated as "human beings", "people", or "persons", he found that "they indicate the position of the subject; they constitute an enunciative marker, not a name" (quoted in Descola 2013: 139). As Descola explains, far from being indicators of ontological exclusion, such ethnonyms simply characterize the point of view of the speaker. In other words, "when they say that nonhumans are persons endowed with a soul, Amerindians are in reality conferring upon them a position as enunciators that defines them as subjects". Viveiros de Castro puts it like this: "Whatever possesses a soul is a subject, and whatever has a soul is *capable of having a point of view*" (quoted in Descola 2013: 139; italics mine). This is philosophically intriguing, since it amounts to *perspectivism* understood as the idea that "any being that occupies a referential point of view, being in the position of subject, sees itself as a member of the human species" (ibid.). Descola's elaboration is worth quoting:

> The human bodily form and human culture are deictics of the same type as ethnonymic self-designations. But that is not to say that perspectivism is a relativism in which each kind of subject forges for itself a different representation of a material world that nevertheless remains identical, since the life of nonhumans is governed by the same values as that of humans: just like humans, nonhumans hunt fish and make war. According to Viveiros de Castro, what are different are the actual things that they perceive: if animals see humans as predator-animals and blood as manioc beer, that is because the

> point of view from which they stand depends on their bodies and their bodies differ from ours in the intrinsic that they manifest. The emphasis that animist ontologies lay on the discontinuity of forms should therefore be regarded as a sign of the heterogeneity of the habitus that a body incorporates as the seat of a particular perspective: "Whatever is activated or 'agented' by the point of view will be a subject" (Viveiros de Castro).
>
> (Descola 2013: 139)

It remains to be seen how far such animist cosmologies take us in the direction of panpsychism, understood as the view that everything that exists possesses mentality and thereby exhibits a point of view in/on the world of which it is part and to which it – somehow, to some degree, high or low, advanced or primitive – actively relates itself, having the (minimum) agency necessary to do so. Perhaps, at the end of the discussion, we will be able to conclude that, as far as anthropological research may help illuminate the issue of panpsychism, animism is the closest we may come to witnessing *panpsychism in practice*, allowing us to explore what a panpsychist doctrine amounts to in the form of a distinct collective practice.

Admittedly, in proposing that animism is panpsychism in practice, I am not presupposing a very precise, very specific definition of panpsychism along the lines discussed – pro et con – in contemporary analytic philosophy à la Nagel. Indeed, the encounter that I set up between philosophical doctrines of panpsychism, on the one hand, and ethnographic studies of animist cultures, on the other, calls for a tentative, exploratory, and work-in-progress-based understanding of panpsychism, concentrating on what is, cosmologically, common among cultures historically and geographically wide apart.

What I propose is common in the sense of a minimally shared common denominator, then, are the very characteristics that Jonas (philosopher, not anthropologist, to be sure) highlights: Panpsychism affirms the life qualities not only of humans but of animals – earth, wind, and water – of the life-sustaining forces operative throughout the natural world, perceived as inseparable from the species-related, goals-realizing, value-producing, and value-sustaining dimension intrinsic to that entire world. As such, panpsychism is the cosmology justified and verified within the available range of experience of a place-bound, local group (culture). To members of such a culture, it "could never occur that life might be a side issue in the universe, and not its pervading rule. Unquestioned and convincing at the beginning stands the experience of the omnipresence of life" (Jonas 2001: 8).

Critical for my purposes in what follows is the intimate way in which "panpsychism in practice", in the form of animism (or what Jonas sometimes refers to as panvitalism), is bound up with *the domain of experience*, with what Jonas designates as "the available range of experience". The in-practice constellation that I am proposing has the form of *a match between thinking and doing, between how the world is being conceived and how it is being experienced and acted in*, in the course of so many encounters and interactions with all the sorts of creatures and forces that go into making it the kind of world that it is. For lack of the required, relevant sorts of – and intact arenas for – such experiences-with, interactions-with, all sorts of nonhuman "others" within a shared world where the respective points of view in/on that world have to be acknowledged and taken into consideration (ultimately, for the sake of survival), the match in question will give way to a mismatch, one both accompanied by and reinforcing a situation of (ecological) imbalance between what are increasingly held to be two wholly separate domains, that of humans and that of nature – which pretty much, however crudely formulated here, catches the trajectory toward "modernity and its discontents", our present predicament.

To show more concretely what my suggestion implies, it is instructive to dwell on a practice that shares a number of features with animism, yet that continues to flourish (albeit to a modest extent) in our otherwise anti-animist society today, namely hunting.

IV

Hunting is a practice whose success depends upon the human hunter's willingness to be committed to a certain set of beliefs about the game – what its properties are – as well as assuming a certain kind of attitude toward the animal in question. To put it somewhat simply: the more the hunter sticks to a purely anthropocentric stance, denying mentality and interiority – agency, intelligence – to the animal, the less likely it is that the hunting will prove a success. Conversely, the more the hunter is willing to adopt the sort of stance toward the animal that animism prescribes – perceiving the animal as a person just like the hunter, to be respected as such, on an equal footing – the greater the chances of success.

Unsurprisingly, Descola discusses hunting in his book. But one of the cases he chooses is unexpected: Aldo Leopold, the forestry engineer trained at Yale who went on to become a founding father of modern American environmental philosophy, largely thanks to his *Sand County Almanac*, published posthumously in 1947 and still a major influence in the field. This pioneer ecologist and ecophilosopher was also, throughout his life, a dedicated hunter, and as Descola remarks he "never suffered from any guilty conscience on that account" (2013: 196). Running through Leopold's reminiscences about his experiences with hunting is one fundamental theme: knowing how to hunt is knowing how to find one's game, and – now comes the truly important part – "knowing how to find it is knowing how to adopt the point of view of the animal that one is seeking, perceiving things as that animal does and putting oneself in its place". It thus involves, Descola continues, "abandoning a superior vantage point in order to seize, from within, upon this tangled web of destinies and desires that weaves together the world in motion". He then provides the wider anthropological context, observing that "such an attitude is reminiscent of the manner in which Amerindian hunting communities envisage the metamorphoses that mark relations between humans and nonhumans, such as exchanges of viewpoints in the course of which each party, modifying the observational position imposed by its body, endeavors to slip into the skin of the other in order to see things from its point of view" (196).

To appreciate what is at issue here, consider Descartes. If Descartes were to act upon his contention that animals are mere "automata", he would make for an altogether inept hunter. Taking himself qua human as the only agent in every sense of the word – intelligence, affectivity, volition, purpose – the animals making up the game would be but a collection of natural objects, passively, mutely, indifferently there for the taking. In other words, he would approach them as devoid of everything it takes to constitute a point of view in/on the world, denying the animals the capacities required for actively partaking in a shared world, in a situation of communication and exchange, of give and take, of call and response, of interpreting what the other's calling and responding means and adjusting one's actions accordingly, as exemplified in the life-and-death decision-making about whether to stay put and keep still, to attack, or to flee.

Descola wisely warns against overstating the analogy between a twentieth-century American ecologist going hunting and a group of Amerindians doing so. He points to the fact that Leopold's celebrated land ethic "in no sense calls into question the ontological distributions of naturalism, which it, on the contrary, accepts without a qualm" (2013: 196). Suffice it to say here that the essence of Leopold's land ethic is captured in the often-cited statement that "A thing is right when it tends to preserve the integrity, stability, and beauty of the biotic community. It is wrong when it tends otherwise" (Leopold 1949: 224f.). To be sure, writes Descola, there is an educational need to "think like a mountain", like Leopold teaches, "the better to evaluate the balance between wolves, deer, and vegetation to be respected on its slopes" (ibid.), this being important to the professional forester and the hobby hunter alike. To Descola, though, it is no more and no less than "a matter of experiencing salutary thoughts of a kind to give substance and a live urgency to abstract ecological learning; in no sense does it constitute a profession of animist faith". Leopold, that is, never does "impute to nonhumans an interiority analogous to that of humans, for the awareness of a future with which he credits them is nothing but a metaphor for the general teleonomy of nature, which, he believes, reverberates within each one of them. Above all, he never ascribes to animals or to plants any ability to lead the existence of a species characterized by cultural conventions, since, for him, the latter are strictly the prerogative of humans" (2013: 196f.). Thus, what Decola finds to be the case with Leopold is precisely what I found to be the case with Serres. For all his celebrated "radical" ecological thinking, Leopold remained loyal to the deeper underpinnings of Western thought about culture and nature in general, as well as to the tenets of the naturalism informing his training as a forestry engineer. Descola is justified in holding that Leopold's adherence to the naturalist schema, including its essential dualism, is borne out in statements such as this: "Wilderness is the raw material out of which man has hammered the artefact called civilization. Wilderness was never a homogeneous raw material. It was very diverse, and the resulting artifacts are very diverse. These differences in the end-product are known as cultures. The rich diversity of the world's cultures reflects a corresponding diversity in the wilds that gave them birth" (Leopold 1949: 188; Descola 2013: 197).

On Descola's analysis, then, in an "ecocentric" cosmology such as advocated by Leopold's most important philosophical follower, J. Baird Callicott, the task of ensuring and preserving its systemic integrity is such as to call for "skilled agents capable of conferring a value upon that which intrinsically has none"; and unsurprisingly, such agents are "recruited from solely among humans". In ascribing to humans and humans alone "the rational interiority and moral conscience" needed to "recognize the essential principle of physical continuity and the material interdependence of all entities in the world", naturalist ontology as we know it and the human exceptionalism that goes with, are confirmed not questioned (Descola 2013: 197f.; Callicott 1989, 1999). On Descola's view, recent developments in some Western countries, such as the concession of a legal personality to domesticated animals, does not call naturalist ontology into question, since the discontinuities of moral faculties remain unchanged. If anything, the extension of the status of a subject to a selected few nonhumans "shows that there was nothing 'natural' about the discrimination of which they used to be objects" (2013: 199). Again, what Descartes postulated as a "timeless" philosophical truth about humans and animals – namely, the ontological abyss between them – is today increasingly recognized as nothing of the sort but instead as a notion produced by a specific mentality born in specific historical circumstances, mainly in sixteenth- and seventeenth-century Europe, a notion henceforth sustained by cultural-symbolic means (narratives) together with scientific and institutional practices as the only "natural" way to conceive of humans and animals alike.

For my purposes, what is most noteworthy here is what Descola calls naturalism's "supreme cunning ploy" in that the allegedly undeniable character of the "self-evident" divisions that it authorizes between the world's various entities stems from the fact that they (the divisions) are said to be founded upon nature. To employ Sartre's distinction once again, the divisions between humans and animals, culture and nature, are said to exist *in re*, so that in acting upon them as the cultural beings that we humans are – and are said to be alone, culture being our prerogative – simply adhering to ("mirroring") the way things are, with special emphasis on how fundamentally – and inalterably – different humans are from nonhumans of all sorts (recall how a twentieth-century Cartesian such as Sartre never bothered to differentiate between nonhuman entities as different as gorillas and rocks; their alleged shared lack of consciousness understood as the capability to negate and transcend any givens sufficed to categorize them ontologically). To claim, on behalf of the ontology within which we Western moderns are socialized, that it is founded upon nature, says Descola, amounts to "an irrefutable argument when it comes to disqualifying rival ontologies" (2013: 199). This is an important observation, because it flies in the face of the notion – consolidated in the social sciences by Max Weber and upheld these days by thinkers such as Habermas (1984, 1987) – that so-called "modern, scientific" world views are self-reflective and in principle falsifiable, in sharp contrast to "mythical" world views – think of Amerindian animism – that insulate themselves against critique and are blind to the extent to which they are man-made, cultural constructions. Although Descola does not say so explicitly, I take him to suggest that the propensity for "naturalization" holds for *all* the major cosmologies that he studies (animism, totemism, analogism, and naturalism) and that, if anything, present-day scientific naturalism is even more stubborn in immunizing itself against exposure to refutation and the threat of being overthrown than are (or have been) its three historical rivals, all of which are presently pushed to the margins due to the all-too-real (as opposed to merely philosophical) hegemony enjoyed by the naturalist outlook and the institutions promoting it on a global scale.

V

There is more to hunting than what Descola's contrast between Leopold and the Amerindians allows us to grasp. I shall draw on the work of fellow anthropologists Eduardo Kohn and Tim Ingold to put hunting into a larger cosmological context, one appropriate to the affinity that I see between animism and panpsychism.

Kohn's *How Forests Think* draws on his ethnographic fieldwork among the Runa of Ecuador's Upper Amazon. It is one of the studies that have contributed most to the renewed interest in animism in recent years.

The Runa are hunters, ineluctably and essentially so, since everything that defines them as a people and culture is inseparable from their hunting practices. Kohn's project is to explore how Amazonians such as the Runa interact with the multitude of creatures that inhabit one of the world's most complex and rich ecosystems. His subject is Amazonian other-than-human encounters, encounters that he draws upon to show how humans' relations – exemplified in Runa's relations – with those beings that exist in some way "beyond the human" force us to question our tidy answers about the human.

Like Karen Barad, Kohn is interested in opening up the field within which encounters take place. Some instances of encounter are more interesting than others – especially inter-, trans-, and cross-species such, turning the tables on the pride of place that the human-to-human case has traditionally been granted in the social sciences. Unlike Barad, however, Kohn is not into quantum mechanics or Bohr-inspired reflections about the knowledgeability of matter as selectively accessed by way of some agential cut. And were he to study the agency of brittlestars, he would probably reach less anthropocentric conclusions than does Barad, notwithstanding her stated posthumanist commitments.

Kohn begins his study on this note:

> How other kinds of beings see us matters. That other kinds of beings see us changes things. If jaguars also represent us – in ways that can matter vitally

> to us – then anthropology cannot limit itself just to exploring how people from different societies might happen to represent them as doing so. Such encounters with other kinds of beings force us to recognize the fact that seeing, representing, and perhaps knowing, even thinking, are not exclusively human affairs.
>
> (Kohn 2013: 1)

Kohn's repeated talk about "going beyond the human" attests to his insistence that "nonhuman life-forms also represents the world". He is quick to note, however, that the expansive understanding of representation that he is propagating is "hard to appreciate because our social theory – whether humanist or posthumanist, structuralist or poststructuralist – conflates representation with language". Setting out to challenge the stubborn tendency to locate signs and their relational and communicational – world-constituting and world-sustaining – role exclusively inside a human framework, Kohn's thesis is that signs also exist well beyond the human. Drawing on work by authors such as Gregory Bateson, Terrence Deacon, and Jesper Hoffmeyer, Kohn's major thesis is that "what differentiates life from the inanimate physical world is that life-forms represent the world in some way or another, and these representations are intrinsic to their being. What we share with nonhuman living creatures, then, is not our embodiment, as certain strains of phenomenological approaches would hold, but the fact that we all live with and through signs" (2013: 9). Building on the semiotic philosophy of Charles Sanders Peirce, yet bringing it to bear on a diversity of other-than-human sign-users not studied by Peirce, Kohn holds that "all life is semiotic and all semiosis is alive". And likewise, echoing Hegel in a context that Hegel would scarcely have anticipated, Kohn contends that "in important ways, life and thought are one and the same: life thinks, thoughts are alive" (16). That the world is "animate" entails that "we" (human persons) are not the only kind of *we* – there are countless others, and should we humans fail to take into account the reality of those other selves and their distinct points of view in/on the world that we share with them, including on us and our actions, we would do so at our peril. As Kohn shows with respect to the Runa, hunting, fishing, and trapping place them in a particular relationship with the many beings that make up the ecology of selves in which they live. These activities force the Runa to "assume their points of view, and indeed to recognize that all these creatures that they hunt, as well as the many other creatures with which those hunted animals relate, have points of view" (17).

The overall point is that different creatures have different points of view; based on their differences in capabilities of perception and action, they attend to the world in different ways. The British anthropologist Tim Ingold, in his seminal *The Perception of the Environment*, observes that Cree hunters, for example, admit that geese (or other animals that they hunt) notice things that humans do not. Cree hunters (of northeastern Canada) acknowledge the importance of attending to the presence of geese "in the knowledge that geese are attending to them" (Ingold 2011a: 51). Indeed, geese are "quite apt at learning in what contexts to expect predation, at learning to distinguish predatory from non-predatory humans, and at communicating appropriate behavioral adaptations to other geese" (52, quoting Colin Scott). Animals participate with humans *qua* persons "as real-word creatures, endowed with powers of feeling and autonomous action, whose characteristic behaviors, temperaments and sensibilities one [say, Cree hunters] gets to know in the very course of one's everyday practical dealings with them" (ibid.). Ingold proceeds to make the important claim that in this regard, "dealing with non-human animals is not fundamentally different from dealing with fellow humans" (ibid.). Accordingly, he contends that Alfred Schütz's definition of sociality, meant as a matter of course for humans only, applies with equal force to the encounter between human hunters and their prey: "Sociality is constituted by communicative acts in which the I [the hunter] turns to the others [the animals], apprehending them as persons who turn to him, and they both know of this fact" (Schütz 1970: 163; Ingold 2011a: 52). And make no mistake: there is no attempt to deny here that there are things only humans can do. As Ingold observes, humans may be unique in their capacity to *narrate* such encounters, to tell stories about them, sometimes based on factual encounters, sometimes being pure fiction that only humans have the required imaginative powers to concoct (we humans tend to believe; and we may or may not be right). However, his point is that "no-one can construct a narrative . . . who is not already situated in the world and thus already caught up in a nexus of relations with both human and non-human constituents of the environment" (ibid.).

Without doubting the validity of Ingold's point, the fact is that, as compared with indigenous peoples, *our* actual participation in the nexus of cross-species relations is rapidly dwindling, weakening by the day: as our (modern, Western) lifeworld is increasingly urbanized, technologized, and as everything to do with the handling of animals, with the provision of what we eat, for its part is professionalized, "outsourced", and taken care of by a minority. Having become out of sight, out of mind, the entire experiential domain of cross-species relations, with all their manifestations of otherness, of difference and variation, fades into oblivion, so that encounters and interaction with *others* is limited almost exclusively to human as opposed to, and to the detriment of, nonhuman others. The result is a tremendous *loss of experiential diversity*, inseparable from the twin-losses of cultural and biological diversity. In this situation, the supposedly unique human capacity of narration that Ingold highlights is eclipsed by another, one perhaps both more definitively uniquely human and easier than ever before to fall prey to: the capacity for denial, as in refusing the truth about the conditions of life upheld in cross-species relations. We are forgetting something that participants of animist life-forms would never permit themselves to doubt: "vital force is free-flowing like the wind, and it is on its uninterrupted circulation that the continuity of the living world depends" (Ingold 2011a: 112).

VI

We shall now explore in more depth the "total environment", to borrow a term from the psychoanalyst Harold Searles (1972), that humans are part of according to nonmodern cosmologies of an animist kind.

What Kohn describes in the case of the Ecuadorian Runa holds for North-American Indian cultures as well. J. Baird Callicott, mentioned above, brings out this similarity when he observes that for the Lakota (Sioux) of the Great Plains, "all living things are kin; they share a common parentage and heritage". Black Elk, a renowned Lakota shaman, talked about "walking the soft earth a relative to all that is". Indeed, the Lakota worldview pictures nature as a large extended family, and therefore "mutual duties and obligations analogous to those governing family relations should also govern human relations with the earth and sky and with all other forms of life" (Callicott 1994: 121). This is a cosmology that locates human beings in a larger *social*, as well as physical, environment; "all creatures, be they elemental, green, finned, winged, or legged, are children of one father and one mother; one blood flows through all; one spirit has divided itself and enlivened all things with a consciousness that is essentially the same", yielding "a cooperative symbiosis of people with their environment" (1989: 189, 190). In his study of American Indians, Callicott warns against designating their posture toward nature as "ecological" or "conservative" in the modern scientific sense of the terms; more to the point, their posture was "moral or ethical" in that their actions manifested "distinctly ethical attitude toward nature and the myriad variety of natural entities", following, Callicott submits, from the Indians view of the natural world as enspirited, which means that "natural beings felt, perceived, deliberated, and responded voluntarily as persons"; and the point about persons is that they are "members of a social order" and as such have the capacity for social interaction (1989: 194).

In saying that the Lakota pictures nature as "a large extended family", Callicott employs a metaphor known from many ethnographical studies carried out by Anglo-American anthropologists. It is worthwhile dwelling on this metaphor since it will help us explore crucial aspects of my suggestion about the affinity between animism and panpsychism that would otherwise go unnoticed.

Tim Ingold offers particularly valuable insights here. I take it that Ingold would object to Callicott's statement on two counts: the Lakota do not "picture" nature, and especially not as "a large extended family". Ingold's argument is a truly philosophical one, and he is making it by reference not to Callicott or the Lakota, but to several well-known ethnographic studies, such as Colin Turnbull's of the Mbuti Pygmies of the Ituri Forest. Indeed, Turnbull's observation is that the Mbuti recognize their dependence on the forest that surrounds them by referring to it as "Father" or "Mother". "They do so", writes Turner, "because . . . it gives them food, warmth, shelter and clothing, just like their parents" (Turnbull 1965: 19; Ingold 2011a: 43). Ingold acknowledges that the analogy is by no means unique to the Mbuti – the analogy, that is, between the most intimate human relations of human kinship and the equally intimate relations between human persons and the nonhuman environment. Many native peoples around the world are known to see themselves as "involved in an intimate relationship of interdependence with the plants, animals and *hala* (including the deities) that inhabit their world" (ibid.). The forest is not something "out there" that responds mechanically or passively but "like a parent" it provides food unconditionally to its "children", the humans.

What prompts Ingold's bigger argument is the common practice among anthropologists in holding that when hunter-gatherers address the forest as their parent, their doing so is "metaphorical": the forest is a metaphor for a (human) parent; addressing the forest as a father or mother is therefore a matter of a human group engaging in projecting something "human" or "social" or "cultural" *onto* something of an ontologically different order, a different domain of reality, namely "nature". Addressed as "father" or "mother", nature (the forest) has been subject to anthropomorphization, whether the humans engaged in it are aware of it or not.

Ingold's plea is that we reverse the order of primacy implied here. To do so, we should simply follow the lead of hunter-gatherers "in taking the human condition to be that of being immersed from the start, like other creatures, in an active, practical and perceptual engagement with constituents of the dwelt-in world" (42), constituents of all different kinds, human and nonhuman, at that. On Ingold's view, then, apprehending the world, attending to the world and everything – everybody – in it, is not a matter of construction (i.e., projection), understood as the work of the mind, but of engagement, "not of building but of dwelling, not of making a view *of* the world but of taking up a view *in* it" (ibid.). The people that the ethnographers study themselves insist that "the real-world landscape in which they move about, set up camp and hunt and gather, is not alien at all but infused with human meaning – that this meaning has not been 'pinned on' but is there to be 'picked up' by those with eyes to see and ears to hear" (57). This is what is meant when they say that they are "at home" in the word: it simply expresses the fact – fundamental and banal at the same time – that the world is "their" world and that it is home. The landscape therefore, on all vital levels – spiritual, historical, personal – is inscribed with the lives and experiences, the triumphs and hardships, of all who through the ages, in that particular region, have dwelt therein, "from distant Time human-animal ancestors to contemporary humans, and *the landscape itself*, rather than anything erected upon it [or mentally projected onto it], stands in memory of these persons and their activities" (54; italics mine). I have italicized "the landscape itself" to highlight that this corresponds with what philosophers designate as realism, including moral realism – only in this case that realism is not one metaphysical world view – or one mental, epistemological model – among others, but is eminently a lived *practice*: picking up and encountering meaning *in re*, as part of the furniture of the world, as inscribed historically into the landscape, so that the landscape is its meaning, the two being two sides of the same coin. Destroy the landscape, and (its) meaning is lost.

Astonishingly, says Ingold, when we turn to the anthropological interpretation of these ways of apprehending the (their) landscape, "we find a complete inversion, such that meanings that the people claim to discover *in* the landscape are attributed to the minds of the people themselves, and are said to be mapped *onto* the landscape". And the latter, "drained of all significance as a prelude to its cultural construction, is reduced to *space*, a vacuum to the plenum of culture" (2011a: 54). Thus the anthropologist F. Myers, for example, writes that "we will consider country *as if it were* simply culturalized space" (quoted by Ingold, ibid.; italics in original). As Ingold points out, the ontological foundation for this interpretative strategy on the part of the (Western-socialized) anthropologist is "an initial separation between human persons, as meaning-makers, and the physical environment as raw material for construction; the 'culturalisation of space' is then what happens when the two are brought into juxtaposition, such that social relations are mapped onto spatial relations" (55). Clearly, this interpretation flatly contradicts the ontology of the native peoples being studied, which is premised on "the fundamental *indissolubility* of the connection between persons and landscape" (ibid.).

For related reasons, Ingold finds fault with the philosophical dimension of Descola's understanding of animism, knowing full well that, based on his field studies with the Achuar of Upper Amazon in the late 1970s, Descola is now widely recognized as the world's leading scholar on animist cosmology. Ingold seizes on Descola's view that, for all the differences between them, both animism and naturalism can be regarded as "mental models which organize the social objectivation of non-humans" (Descola 1996: 87; Ingold 2011a: 107). Ingold's complaint is that "this appeal to the language of mental models, to the idea of accommodating beings that are really non-human into schemes of representation that construct them as social and therefore human, belongs squarely within a naturalist ontology, and it is from this that the terms of the comparison are derived". What the terms that Descola employs in fact do, according to Ingold, is "to preserve a space for 'really natural' nature which is unaffected by the diverse constructions that the human mind might place upon it. Thus the comparison between naturalism and animism, since it is done on naturalism's terms, is hardly a fair or balanced one" (Ingold 2011a: 107). The upshot is that in animic systems as accounted for by Descola, the powers of intentionality and awareness that we normally attribute to persons are being projected onto nonhuman kinds. And if we follow Descola in "assuming that in reality [such powers] are reserved for human beings, such projection is bound to be anthropomorphic". In other words, "if only humans *really* have intentions" – and this is the *petitio principii* that Ingold sees at work in Descola – "to represent non-humans such as bears *as though* they were persons with intentions is necessarily to represent them as human". For this reason, Ingold charges Descola with building a component of anthropomorhism into his very definition of animism, portraying it as "a system that endows natural beings with human capacities" (ibid.). In doing so, animists (to put it in my terms, not

Ingold's) commit a fundamental category mistake: they systematically get the entities in the world wrong, mistaking the essentially and irreducibly nonhuman for the human.

I am not convinced that Ingold's criticism is borne out in Descola's *Beyond Nature and Culture*; his earlier work may represent a better target. True, at one point in his book Descola states that animism and naturalism "set up human society as the paradigm of collectives" (2013: 267; not quoted by Ingold), and in a footnote he mentions, without comment, that Viveiros de Castro for his part "describes animism as anthropomorphic, in contrast to the narcissistic anthropocentrism of Western evolutionism" (de Castro quoted in Descola 2013: 423, fn. 14). Yet he also makes the somewhat ambiguous remark that "animism is anthropogenic rather than anthropocentric in that it derives from humans all that is necessary to make it possible for nonhumans to be treated as humans" (2013: 258).

My concern here is not to settle the dispute between Ingold and Descola but to let it illuminate the distinction between de-anthropomorphization and anthropomorphization that is crucial to Ingold's argument. He cites the common view that natural science seeks to describe the world through a "progressive de-anthropomorphization", that is to say, through the attempt to "expunge from its notion of reality all that can be put down to human experience. Thus purified, nature is revealed to a detached human reason as a domain of things in themselves" (Ingold 2011a: 108); nature as wholly and purely *en-soi*, as it were, without the mediation and so (presumably) bias of the *pour-soi*, to invoke the distinction known from Sartre and employed by Griffin and discussed in Chapter 2. Thus conceived, I suggest that de-anthropomorphization would be tantamount to Nagel's view from nowhere.

The point Ingold goes on to make is this: animist ontology, too, entails a process of de-anthropomorphization, only it "operates in a quite different direction: instead of severing the link between reality and human experience, [animist] ontology recognizes *the reality of the experience of other-than-human-beings*" (2011a: 108; italics mine). In doing so, I propose, this ontology articulates the core of a panpsychic outlook on reality, as defined in earlier chapters. That the recognition thus articulated cannot be forthcoming from "nowhere" but can only do so from some *particular* point of view – meaning something's or (more to the point) *someone's* such, in keeping with the Gadamerian insight I invoked above – is well understood among practitioners of animism; indeed, I take it to be one of that cosmology's key insights. Ingold puts it like this: "All experience depends on taking up a position in the world, tied to a particular form of life, but for the Ojibwa [indigenous hunters and trappers of the forests to the east of Lake Winnipeg and north of Lake Superior in Canada] the human is but one form out of many". This leads Ingold to reiterate his criticism of Descola: it undermines the latter's characterization of animist systems as inherently anthropomorphic, following the assumption that "experience depends upon powers of awareness and intentionality that mark their possessors as uniquely human" (Descola 2013: 108). It is as if Descola the French anthropologist has de-exotized the alien (i.e., animist ontology as compared with naturalism) by depicting its epistemological underpinnings through the lens of the all-too-familiar, failing to question his own tacit assumption (being that of his native naturalism modern Western style) that humans are the only "real" and "true" possessors of experience, that is, of awareness, of mentality, of powers to initiate and respond to actions, to take part in communication, and the like, and so unique in being able to constitute – indeed, actively adopt and act upon – a point of view on/in the world. Again, this strikes me as an exceedingly strong, and possibly mistaken, critique of Descola, yet I elucidate it here because it serves to bring out the overall argument and the rival positions involved.

The philosophical implications are set out by Ingold in a passage worth quoting in full:

> In the economy of knowledge of the people themselves, there are not two worlds, of nature and society, but just one, saturated with personal powers, and embracing both humans, the animals and plants on which they depend, and the features of the landscape in which they live and move. Within this one world, humans figure not as composites of body and mind but as undivided beings, "organism-persons", relating as such both to other humans and to non-human agencies and entities in their environment. Between these spheres of involvement, there is no *absolute* separation, they are but contextually delimited segments of a single field. [. . .] To coin a term, the constitutive quality of their world is not intersubjectivity [considered to be the constitutive quality of the social domain *as against* the object world of nature, a domain open to human beings but not to non-human kinds] but *interagentivity*. To speak of the forest as a parent is not, then, to model object relations in terms of primary intersubjectivity, but to recognize that at root, the

> constitutive quality of intimate relations with non-human and human components of the environment is one and the same.
>
> (Ingold 2011a: 47)

Based on this analysis, Ingold proposes the following understanding of life:

> To be alive is to be situated within a field of relations which, as it unfolds, actively and ceaselessly brings form into being. Humans as humans, geese as geese, and so on. Far from revealing forms that are already specified, life is the process of their ongoing generation. Every living being, then, emerges as a particular, positioned embodiment of this generative potential. Hence personhood, far from being "added on" to the living organism, is implicated in the very condition of being alive: the Cree word for "persons" . . . can itself be glossed as "he lives". Organisms are not just *like* persons, they *are* persons. Likewise, consciousness is not supplementary to organic life but is, so to speak, its advancing front – "on the verge of unfolding events, of continuous birth", as [Colin] Scott renders the Cree conception.
>
> (Ingold 2011a: 51)

VII

It will not be lost on the reader that there are several occasions for recalling Whitehead in Ingold's analysis. For example, the notion that the landscape is "really", in itself, a tabula rasa, devoid of meaning as well as value and so passively awaiting that meaning – and the whole series of properties associated with it – be "pinned on" it, projected onto it, by the only agents seen as capable of doing so, i.e., humans, echoes Whitehead's discussion of the fallacy of misplaced concreteness, though obviously actualized here in a different context than the one Whitehead had in mind. Likewise, Ingold's suggestion that "consciousness is not supplementary to organic life but its advancing front", is certainly in tune with Whitehead's philosophy of organism in *Process and Reality*. In fact, Ingold himself makes the reference in the one quotation from Whitehead that he brings in his book: what we call mind, says Ingold, "is the cutting edge of the life process itself, the ever-moving front of what Whitehead called a 'creative advance into novelty'" (Whitehead 1929: 314; Ingold 2011a: 19).

One of the most valuable insights in Ingold's work concerns the way he links his understanding of animism with a particular view of materiality. Rather than speaking, generally and metaphysically as philosophers are wont to do, of the material world, asking what its intrinsic properties are, we should speak of "the world of materials" (2011b: 31). Ingold attacks "the misrecognition of the active properties of materials as the power of the materiality of objects": "Far from being the inanimate stuff typically envisioned by modern thought, materials in the original sense are the active constituents of a world-in-formation. Wherever life is going on, they are relentlessly on the move – flowing, scraping, mixing and mutating. The existence of all living organisms is caught up in this ceaseless respiratory and metabolic interchange between their bodily substances and the fluxes of the medium" (2011b: 28). On Ingold's view, *things are in life rather than life in things*: thus the connection he maintains between animism and materiality may be captured. Again, his point is that animism is profoundly distorted when understood as entailing the (human-made) attribution of life, spirit, or agency to objects (or materials, or – unspecified – matter as such) that are "really" inert, thus rendering the animating principle as something purely *additional* to the material object on which it has been bestowed (whether that act of bestowing is acknowledged or not by the human bestowers in question). Here is the genuinely animist take on materiality as Ingold sees it:

> In the world [of peoples committed to animist cosmology] there are no objects as such. Things are alive and active not because they are possessed of spirit – whether *in* or *of* matter – but because the substances of which they are comprised continue to be swept up in circulations of the surrounding media that alternately portend their dissolution or – characteristically with animate beings – ensure their regeneration. Spirit is the regenerative power of these circulatory flows which, in living organisms, are bound into tightly woven bundles of tissues of extraordinary complexity. All organisms are bundles of this kind.
>
> (2011b: 29)

This in turn brings Ingold to the following understanding of animism, stated most cogently in his work *Being Alive*:

> In the case of animism, we are dealing not with a way of believing *about* the world, but with a condition of being *in* it. [. . .] Animacy is not a property of

> persons imaginatively projected onto the things with which they perceive themselves to be surrounded. Rather it is the dynamic, transformative potential of the entire field of relations within which beings of all kinds, more or less person-like or thing-like, continually and reciprocally bring one another into existence. The animacy of the lifeworld, in short, is not the result of an infusion of spirit into substance, or of agency into materiality, but is rather ontologically prior to their differentiation.
>
> (2011b: 67f.)

The view Ingold tirelessly argues against, then, is what he calls "the logic of inversion", a logic at work not only in Western metaphysics since Descartes (the usual culprit and fall guy) but also, more surprisingly, in the studies carried out by leading anthropologists, showing, one is tempted to suggest, just how recalcitrant is the legacy of the former also within academic disciplines that, as far as their subject matter is concerned – mostly, "nonmodern" cultures and cosmologies – could be expected to have liberated themselves from it. The logic of inversion that Ingold has in mind is one that "places occupation before habitation, movement *across* before movement *through*, surface before medium". This logic is intellectually incapable of doing justice to the animist ontology where "what is unthinkable is the very idea that life is played out upon the inanimate surface of a ready-made world". Taking *winds* as his example, in most animist cosmologies they are perceived to be alive and to have agentive powers of their own. To make sense of this, insists Ingold, we have to recognize the primacy of movement in the animist cosmos: that things are never static, never simply "what they are" prior to and independent of the relationships with other things – "things" being admittedly my term, not Ingold's, yet it is one that I use deliberately here to capture that the observation just made applies to *everything* that is part of the world as experienced by animist peoples. "We are not required", writes Ingold, "to believe that the wind is a being that blows, or that thunder is a being that claps. Rather the wind *is* blowing, and the thunder *is* clapping, just as organisms and persons *are* living in the ways peculiar to each" (2011b: 73).

VIII

Given this cosmology, how can you account for the practice, among Lakota and Runa alike, of hunting, of deliberately killing animals deemed as persons and experienced as the siblings of humans? Is there a contradiction here? If so, for "us" or for "them", or in both cases?

Callicott remarks, matter of factly, that traditional American Indians seem to have had "no compunction about taking the lives of animals and plants", provided that the hunter in every case acknowledges his prey as persons and pays them the respect due to all persons, regardless of species and their different modes of locomotion. In fact, Callicott views the kinship terms used to address animals and plants as evidence that the Indians "regarded nonhuman life-forms as, if anything, superior to themselves". Indeed, he continues, animals and plants, "in permitting themselves to be taken for legitimate human needs [ultimately, survival], are said to 'pity' people and to voluntarily sacrifice themselves for the sake of their younger siblings, the human beings" (1994: 124). In this way, we may add, nonhuman species are indispensable – and thus factually *not* a mere, exchangeable means – to humans in a twofold sense: first, for human existence to persist physically and biologically, and second, for reminding humans about the basic fact of the *condition humana*: humans are the inferior party in the overall inter-species inter-relationship in that they depend on the gift of nonhuman species, on their never-ending generosity in helping sustain human existence on earth.

However, the present-day American environmental philosopher Callicott tends to downplay the ways in which cross-species relationships, as acted upon in animist cultures, involves conflict and struggle, both philosophically and pragmatically. The anthropologist Kohn has no qualms about addressing this head-on: when the Runa enter the forest's ecology of selves in order to hunt, they "recognize others as selves like themselves in order to turn them into nonselves" (Kohn 2013: 17). Tellingly, the concept "nonselves" never turns up in Callicott's discussion. In Kohn's terminology, turning other selves into nonselves amounts to objectification, the necessity of doing so being, he says, the flipside of animism, and as such "not a straightforward process". He goes on to observe that "one's ability to destroy other selves rests on and also highlights the fact that one is an ephemeral self – a self that can all too quickly cease being a self". Kohn uses the term "soul blindness" to probe the situations where the ability to recognize other selves (i.e., those of nonhumans) is lost, "resulting in a sort of 'monadic alienation' as one is, as a consequence, avulsed from the relational ecology of selves that constitute the cosmos" (ibid.).

Whereas Hans Jonas complained about the conspicuous absence of death in panpsychism à la Whitehead, and Callicott for his part fails to explore the peculiar relationship between life and death in Indian animist cosmologies, Kohn coolly observes that death is intrinsic to life, making it a "difficulty of reality", a "fundamental contradiction that can overwhelm us with its incomprehensibility". There is a feeling of "disjunction" in the Runa who do their hunting in a vast ecology of selves and who "must stand as a self in relation to so many other kinds of selves who one then tries to kill"; indeed, the practice of hunting highlights that "the entire cosmos reverberates with the contradictions intrinsic to life" (17f.).

Whereas Kohn's emphasis on contradiction in the sense given is well taken, especially in correcting Callicott's account, it does not fully capture the cosmological significance that hunting possesses in an animist ontology. The premise on which the practice of hunting is predicated here is that beings of both human and nonhuman kinds are engaged in ongoing mutual interaction, of which hunting and the roles therein distributed are an integral part. Hunting secures that there is a balance in the give and take of life and death, vitality and mortality, a balance based upon as well as sustaining the dependency of all involved, humans and nonhumans alike. Ingold articulates this point better than both Kohn and Callicott, observing that "hunting effects the circulation of vital force between humans and animals and thus contributes directly to the regeneration of the lifeworld of which both are part. The animals offer something of their potentiality and substance to human beings so that the latter may live, while humans, in return, through the proper treatment of animals in death, ensure the release of their life force and hence their subsequent reincarnation". Hence human life is "predicated on the mortality of animals" (2011a: 114; cf. Berkes 2008: 103ff.). Responding to Descola's remark that hunting is "a quite mundane activity of food procurement" (1996: 95), Ingold, characteristically, stresses its cosmological significance: "In the animist ontology, the killing and eating of game is far more than mere provisioning; it is world-renewing" (2011a: 114).

The principal point that Ingold is making, and the one most relevant to my thesis about the affinity between panpsychism and animism, is that as long as we continue to assume – as indeed even leading anthropologists do, if we are to believe Ingold – that only humans can truly be persons, the attribution of personhood to animals is bound to be anthropomorphic: a matter of (mistakenly) ascribing human qualities to nonhuman beings (see 2011a: 91). Now the crucial point about an animist ontology such as practiced by, for example, the Ojibwa, is that persons can take a great variety of forms, of which the human is just one. Again, this is not an "idea", let alone a hypothesis, that they have; it is a matter of *what they experience in their world* and in that sense both presuppose as being true and help sustain as true, by way of acting in the appropriate manner. Persons, that is to say, can appear in a variety of animal guises; they may also appear "as meteorological phenomena such as thunder or the winds, as heavenly bodies such as the sun, and even as tangible objects such as stones that we would have no hesitation in regarding as inanimate" (2011a: 91). Note that "none of these manifold forms in which persons appear is more basic, or 'literal', than the others" (ibid.). And most importantly, persons can change their form. Indeed, for a people like the Ojibwa the capacity for *metamorphosis* is one of the key aspects of being a person, and it is "a critical index of power: the more powerful the person, the more readily a change of form may be effected" (ibid.). Accordingly, "only the most powerful human persons, such as sorcerers and shamans, can change into a non-human form and make it back again – and then only with some danger and difficulty. Sorcerers, for example, can transform themselves into bears in order better to pursue their nefarious activities" (2011a: 93).

What can we learn from Ingold's account? I propose putting it like this. Personhood in animist ontology is not a case of "anything goes", i.e., that nothing is, or need to be, fundamentally similar in all the different beings and entities – even counting inanimate ones such as stones – that qualify as persons, and hence are concretely perceived and treated accordingly. As acted upon, and in that sense as panpsychism in practice, the point is that everybody – human agents among them – act and respond to all encountered "others" so as to help ensure that everybody/everything that is a person is allowed to sustain himself/herself/itself in that kind of being. And – this being the bottom line – to be sustained is not (never, for no one) a matter of individual self-sufficiency but requires the participation and support of innumerous others, including very different sorts of persons/creatures from one's own. In other words, sustenance of being a person is a truly cross-species undertaking, a truly joint effort and attainment if there ever was one, and as such a holistic and ecological one.

So what, then, on this animist outlook, is similar to all persons, similar in a sense not contradicting or incompatible with their said differences across the vast cross-species spectrum and continuum? Ingold is adamant that "all persons, whether human or not, share the same fundamental structure" (2011a: 92). He follows the leading anthropological authority on the Ojibwa, A. Irving Hallowell, in holding this structure to consist of "an inner vital part that is enduring and an outward form which can change". The inner essence, or soul, explains Ingold, "holds the attributes of sentience, volition, memory and speech. Any being that possesses these attributes is a person, irrespective of the intrinsically unstable form in which it appears" (ibid.). And "while human persons and other-than-human grandfathers are alike in this regard, such that no absolute division in kind can be drawn between them, they do differ in degree – that is, in the amount of power a person possesses and hence in their capacity for metamorphosis. Grandfathers are more powerful than living humans. Most powerful are the Sun, the Four Winds, the Thunder Birds, and the spirit 'masters' of all the different species of animals. These beings are immortal, but can change their form with relative ease, appearing now as a human, now as an animal, now perhaps as some meteorological phenomenon" (2011a: 93).

Given that the animist cosmos is populated by beings of both human and non-human kinds, all engaged in ongoing mutual interaction, we are reminded of Karen Barad's notion of agency as a capacity found in all sorts of organisms, from the human person to the brittlestar; indeed, we saw above that Ingold uses the term "interagentivity".

I have two comments as regards to comparing the two. First, to recall my earlier discussion, one of the costs of Barad's approach is that in granting agency to virtually "everyone", finding it virtually "everywhere", depending on the access gained according to the "cut" in each case, the differences – involving inequalities of power – between the manifold of beings said to share the same basic capacity (agency) are not adequately accounted for; hence Barad's is a case of "flat ontology". It is as though highlighting differences would disturb the egalitarianism – as between all species, all beings – that posthumanism champions. Yet a true egalitarianism, one worth advocating, would be one that celebrates differences rather than downplaying them. Second, though animist ontology shares Barad's non-essentialism, her rejection of properties as "innate", given, and unalterable, and her openness to processes of change and everyone's part in those, her agential realism stops short of grasping *change* in the radical sense of metamorphosis that is so characteristic of animist ontology. As Ingold explains, there is no absolute distinction between "visiting the other" and "becoming the other", since such becoming is not a permanent phenomenon but a temporary one, thus allowing for a "going back" to what and where one came from. Since animals, as well as humans, form their own communities, members of each can – and do, as borne out in concrete encounters as well as in dreams (whose reality claim is denied or neglected only at one's peril) – visit the communities of the other. Hence, upon "crossing over" to the animal side, a man "will see his hosts as creatures like himself, while to the people back home he will now appear in animal form". In short, "in the dialogue between human and animal, each in turn takes up the point of view of the other, becoming temporarily other to his or her own people" (2011a: 114). As mentioned above, the shaman is a person of exceptional power in that he can move with relative ease across the human-animal interface. The point is that this holds for animals as well; not all of them, but some. Particular animals are credited with powers similar to that of the shaman; Ingold observes that the bear is a case in point right across the circumpolar North. "If the bear can appear human", writes Ingold, "so too, the human shaman can show up as a bear. Whether you see one or the other depends on where you are looking from; in other words it has to do not with the substance of being but with the relative positioning of self and other in contexts of dialogue" (2011a: 114).

IX

Against this background, we shall now look more closely at Kohn's work, in particular his bold claim that forests think. How can he make such a claim? Why does he not limit himself to studying how humans (say, Runa) think about their forest, or – if he insists on going beyond that – how people such as the Runa think forests think?

But no. Kohn wants to be radical: his stated aim is to "show that the fact that we can make the claim that forests think is in a strange way a product of the fact that forests think". The two, he says, are related: "It is because thought extends beyond the human that we can think beyond the human" (22). In making the case for this claim, Kohn draws on Peirce's semiotic philosophy, especially his understanding of a sign as "something which stands to somebody for something in some respect or capacity" (Peirce CP 2.228, quoted in Kohn 2013: 29). The point is that Peirce's notion of the sign breaks with the dominant Saussure-inspired view that human language is the paragon and model for all sign systems. By contrast, Peirce's definition of the sign is "agnostic" about what signs are and (not least importantly) "what kinds of beings use them"; for Peirce "not all signs have language-like properties and not all the beings who use them are human". His broader definition of the sign helps us "become attuned to the life signs have beyond the human as we know it" (29). It follows that, while signs always mediate, i.e., they help transmit information; they do not necessarily exist in some separate domain inside (human) minds and cut off from the entities they "stand" for. In other word, signs are not just *about* the world, as a simple, tacitly anthropocentric, representationalist epistemological model would have it. "Signs are also in important ways *in* it" (30; italics mine).

To explain what is at issue here, Kohn invokes an incident that he witnessed while joining two Runa for hunting. Kohn, together with a man named Hilario and his son Lucio, came upon a troop of wooly monkeys moving through the canopy. Lucio shot and killed one of the monkeys; the rest dispersed. However, they noticed that one young monkey became separated from the troop; being alone, she hid in the branches of an enormous red-trunked tree that poked out of the forest canopy high above. This is what happened:

> In the hope of startling the monkey into moving to a more visible perch so that his son could shoot it Hilario decided to fell a nearby palm tree. [. . .] As Hilario had anticipated, the sound of the palm tree crashing frightened the monkey from her perch. This event itself, and not just the before-the-fact imitation, can also be taken as a sign. It is a sign in the sense that it too came to be "something which stands to somebody for something in some respect or capacity" [Peirce]. In this case the "somebody" to whom this sign stands is not human. The palm crashing down stands for something to the monkey. Significance is not the exclusive province of humans because we are not the only ones who interpret signs. That other kinds of beings use signs is one example of the ways in which representation exists in the world beyond human minds and human systems of meaning.
>
> (Kohn 2013: 31)

The question is: what does the crashing palm tree come to signify for the monkey? Kohn's answer is that the crash, as a sign in Peirce's sense, is not a likeness of the object that it represents; instead, it points to something else, namely the sort of sign Peirce calls an "index", constituting his second broad class of signs. Whereas iconic and indexical modes of reference form the basis for all representation in the living world in which human hunters and monkeys alike actively participate, Peirce's third kind of sign, "symbol" (as in symbolic reference) is, as far as we know, a form of representation that is unique to humans. This is because symbols refer, not simply though the similarity of icons, or solely through the pointing of indices, but "to their object indirectly by virtue of the ways in which they relate systematically to other such symbols"; and symbols involve convention (32).

If this gets technical, consider the example with the monkey. As Kohn observes, although we might suggest that the hunter's (Hilario's) tug, propagated through the liana and mat, literally shakes – and frightens – the hiding monkey out of her sense of security, "how this monkey comes to take this tug as a sign cannot be reduced to a deterministic chain of causes and effects". Why is that? The monkey, Kohn submits, "need not necessarily perceive the shaking perch to be a sign of anything", anything in particular, I would like to add. "And in the event that she does", Kohn continues, "her reaction will be something other than the effect of the force of the tug propagated up the length of the liana" (32). Here is the decisive point: If indices, as Peirce holds, involve something more than mechanical efficiency, that something is, paradoxically, something less: it is an absence. Kohn explains: "To the extent that indices are noticed [be it by a monkey] they impel their interpreters to make connections between some event and another potential one *that has not yet occurred*" (32). To the hiding monkey, that is, the moving perch – the fact that some agent, for some reason, has brought about its moving – is a sign, and as such is connected to something else: that for which the movement stands, i.e., its meaning and significance as something happening in *that* situation, one in which hunters have brought the monkey into hiding, realizing that it is prey and hence in danger. What is certain, then, is that "something is about to happen, and she [the monkey] had better do something about it". This underscores Kohn's principal point: What indices do is that they "provide information about such absent futures. They encourage us to make a connection between what is happening and what might potentially happen" (33), "us" being precisely not an exclusively human agent but animals such as monkeys as well. When the hunter fells the tree and causes it to come crashing down, both the hunter and the monkey know that the monkey will respond in kind (by fleeing, going into hiding). In Kohn's formulation, such tropical trans-species instances of communication "reveal the living worldly nature of semiosis"; "all semiosis", he continues, "(and by extension thought) takes place in minds-in-the world" (34).

Drawing on the work of Terrence Deacon, Kohn holds that "constitutive absence" is a "special kind of nothingness"; significantly, the categories of absence and nothingness are not just found in the world of artifacts or humans. According to Deacon, constitutive absence is "a kind of relation to that which is spatially or temporally not present that is crucial to biology and to any kind of self" (Deacon

2012: 3). Kohn cites the quote Deacon gives from Gregory Bateson: "In the world of mind, nothing – that which is *not* – can be a cause" (Bateson 1972: 452). The general point endorsed by all three is that in the "world of mind" constitutive absence (in the words of Kohn) is "a particular mediated way in which an absent future comes to affect the present. That is why it is appropriate to consider telos – the future for the sake of which something in the present exists – as a real causal modality wherever there is life" (Kohn 2013: 37), this being a major thesis argued in Deacon's widely discussed *Incomplete Nature*.

As we saw in the example brought by Kohn, the notion of constitutive absence applies to the animal prey as well as to the human hunter, as it were, in that the monkey took the latter's felling of the palm tree as a sign in Peirce's sense. The felling, being a deliberate action, is a signal to the monkey that *something is about to happen*: the felling is expected to be followed by further action; it is *meant to . . .*, it is something done *in order to . . .* ; and yet, at the moment the monkey responds "in kind", as I said, the "something" about to happen has not happened (yet); as such, at the time the monkey responds-by-fleeing/hiding, it is a nothingness, an absence. And of course – it now becomes irresistible to point out – should the monkey wait until that nothing (a futural such, as sheer potentiality) has turned into a definite something, it will be *too late*: staying put instead of fleeing, the monkey may expose itself so as to be shot on the spot. In other words, the monkey's ability to react promptly, unhesitatingly, *prior* to the manifestation of the hunter's follow-up, his planned next move, and in that sense to respond to an absence, a not-yet, the possibility that . . ., is a capacity without which it would not be able to live – survive – in the forest, given the dangers to it represented by the aims of other creatures in it, such as predators, be they human or animal.

It is interesting to note – none of the authors mentioned do – how flatly this thesis contradicts what for so long has been the dominant view in philosophy in general and in the philosophy of mind in particular. Sartre is, again, an instructive case in point. To Sartre, it goes without saying that consciousness (his preferred term over mind) is the unique prerogative of humans. What is most telling in the present context, though, is his obsessive emphasis on absence and nothingness: the for-itself, *pour-soi* (Sartre's technical term for consciousness), is "that which it is not and is not that which it is"; whereas the in-itself, *en-soi*, is "that which it is", period, meaning it is pure, and indifferent, self-identity (Sartre 1956). The upshot is that only the for-itself (that is, humans) is capable of negation and of transcendence: of going beyond any given state of affairs in the world so as to assess it, and subsequently act upon it, as *lacking* and as either opening or preventing certain "projects" being adopted and pursued, certain goals of action being reached. To judge that something that is is somehow "lacking", presupposes the ability to conceive – imagine, project – what it could be, if only . . .; and forming such a conception involves *value*, in Sartre's sense, because it depends on the assessment that something otherwise than what is would be, for some reason and in some particular respect(s), given one's adopted project, *better*, preferable, and so worth pursuing in one's choice of action.

There is an affinity worth mentioning here between Sartre's understanding of consciousness and Hannah Arendt's (1978) definition of thinking; two thinkers whose convergence on this matter is all the more noteworthy given the stark differences routinely noted between them: to think is to be able to imagine as absent what is in fact present, and as present what is in fact absent. In Arendt's case no less than in Sartre's, it goes without saying that thinking, thus understood, separates humans from nonhumans in a most unequivocal and indisputable manner. In particular, thinking epitomizes what Arendt calls "the life of the mind" in that it is only possible for beings capable of liberating themselves, in their reactions to things in their surroundings, from the determinate and as it were reaction-dictating power of the senses. Inasmuch as the senses are taken by Arendt to only, more or less passively, provide information about what is the case in the form of being-there, being present, manifest, constituting a fact-in-the-world as accessed from a particular here-and-now, it falls to thinking to effect the breach (through negation) with the given that Sartre for his part deems the outstanding feature of consciousness.

So far, so good. But I have a critical question about Kohn's endorsement of Bateson and (more importantly) Deacon that will prove important for what follows. Could it be that they their way of theorizing absence, of granting it a place in the wider-than-human world, is in fact considerably less groundbreaking than Kohn invites his readers to think?

Consider, first, Bateson. Recall the statement of his cited by Deacon: "In the world of mind, nothing – that which is *not* – can be a cause" (Bateson 1972: 452). To explain his idea, Bateson gives two examples: "The letter which you do not write can get an angry reply; and the income tax form which you do not fill in can trigger the International Revue boys into energetic action" (ibid.). True enough. Sartre would certainly agree; in fact, both examples could easily have been found in his own work. Those – the agents – for whom nothing can be a cause, are taken to be as human agents, such as Kohn's monkey, are mentioned. Bateson's idea, at this juncture at least, is wholly compatible with an anthropocentric outlook, although, in his later work *Mind and Nature*, Bateson clearly aims to go beyond it.

Be that as it may – the case of Deacon is surely of much greater systematic importance for Kohn's argument. Since Deacon's *Incomplete Nature* is widely hailed as an intellectual *tour de force*, I shall subject his position to closer scrutiny, before returning to Kohn's work and its relevance for my suggestion that animism be considered as panpsychism in practice.

X

Deacon begins his book with the proposal we saw Kohn citing: the term *absential* is meant to "denote phenomena whose existence is determined with respect to an essential absence" (Deacon 2012: 3). This is no doubt intriguing, and it immediately sparks questions such as this: for *what kinds of beings* does it hold that their "existence is determined with respect to an essential absence"? What does it take, in terms of agential capacities and powers, to *have absence* (or what is deemed, from the point of view in question, instances of absence) *matter*? To paraphrase Bateson, for whom and under what circumstances are absence constituted as "a difference that makes a difference"?

In the context of Kohn's study of how it can be that "forests think", which indeed captures his principal thesis, Deacon's answer to these questions must surely be instructive. As far as I can see, the answer is this: "This paradoxical intrinsic quality of existing with respect to something missing, separate, and possibly nonexistent is irrelevant when it comes to inanimate things, but *it is a defining property of life and mind*" (2012: 3; italics in original). He goes on to say that "a complete theory of the world that includes us, and our experience of the world, must make sense of the way that we are shaped by and emerge from such specific absences". So, while it is beyond doubt that "what is absent matters", the fact is, and very much remains, that "our current understanding of the physical universe suggests that it should not"; hence, "a causal role for absence seems to be absent from the natural sciences" (ibid.).

Far from finding this clarifying, I am confused. The phenomenon of absence, and the whole series of questions it raises, the consequences it will have, etc., are said – without argument – to be "irrelevant when it comes to inanimate things", being, in contrast, "a defining property of life and mind" (2012: 3). Again, Sartre would be in complete agreement with the first part (inanimate things being in-itself and as such alien territory, the wrong doors to knock at, as far as absence and its mattering are concerned). He would be less sure about concurring with the second part; it all depends on how life and mind are defined and what sorts of beings are held to exemplify them.

To take stock, we note that Deacon is happy to invoke, without argument, the orthodox dualism between inanimate and animate (without further specification) in making his claim about the importance of absence: absence is said to be absent in the inanimate domain, present – making a difference – only in the animate (details not yet given). The problem as Deacon sees it, though, is that "our current understanding of the physical universe suggests that [absence] should not [matter]", the fact being, however, that it does. But then again, *for and among whom does it matter*, on his thesis? We are not told. We only know the negative answer: not for and among inanimate things.

Deacon prepares the ground for his case for the importance of absence by offering a broad historical narrative according to which "absential accounts of all kinds came into question" following the mechanistic accounts of inorganic phenomena that goes back to Descartes and Hobbes, at least. And then, in 1859 with Darwin's notion of natural selection, "even the special order of living design seemed to succumb to a non-absential account". More recently still, in the latter half of the twentieth century, with the study of so-called self-organizing processes (i.e., autopoiesis), it became "even more common to hear of absential accounts described as historical anachronisms and illusions of a prescientific era". Indeed, Deacon goes on to say, "many scholars now believe that developing a science capable of accurately characterizing complex self-organizing phenomena will be sufficient to finally describe organic and mental relationships in entirely non-absential terms" (Deacon 2012: 4f.).

It would take us too far afield to give an assessment of Deacon's attempt to do justice to the importance of absence as he understands it, thereby taking himself to go against important strands of current thinking such as those just mentioned. Suffice it here to recap my question: for and among whom does absence matter, on Deacon's thesis? In a section entitled "Omissions, expectations, and absences" (sounding very much like Sartre's discussion in *Being and Nothingness*, a work never mentioned by Deacon), he brings some examples that allow us to try and answer my question. Deacon observes that "the importance of physical regularity to the analysis of information is that it is ultimately what enables absence to matter. Where something tends to regularly occur, failing to occur or in other ways diverging from this regularity can stand out" (2012: 376). This seems true, as does his further observation that "a predictable behavior [. . .] is presupposed in many circumstances where something absent makes a difference. The source of this regularity *need not be exemplified by human action*" (ibid.; italics mine). Again, well taken. And Deacon's example to drive home the point? It is this: "It could, for example, be exemplified by the operation of a machine, as when stepping on the brakes fails to slow one's vehicle and thus provides information about change" (ibid.).

As for providing an answer to my question, this is less than satisfactory, to put it mildly. True, the "absence" that figures in the example is one showing up in a machine, in a human-made artifact; it takes the form of the driver's stepping on the brakes failing to have the expected (and regular) effect, namely slowing the car. Considered as an example that would help illuminate what is philosophically important about absence and the role it plays for *various kinds of beings (agents)*, for whom absence matters, it is a non-starter. Deacon's example may locate the absence in the machine. But everything to do with noticing that absence, in its simple form as a failure to behave as expected and proscribed for the machine that it is, expected and proscribed by human agents and no others at that, is located purely within the human agent taking part in the situation, i.e., the driver insofar as the brakes fail to help him reach his desired goal – stopping the car. Remove the driver and his goal in stepping on the brakes, and there would be no absence to talk about, no absence that would matter, in the example. Thus my question about the "for and among whom" has been answered: for a human agent. This being so, nothing in terms of going beyond the orthodoxy of dualism and anthropocentricity (Kohn's project, we recall) has been gained.

To be fair, this example does not exhaust Deacon's repertoire. He also pays attention to the realm of social interactions, in which we are "quite familiar with circumstances where omissions can have major consequences", for example, "failing to submit a U.S tax return by April 15", this being, incidentally or not, one of the examples that we saw offered by Bateson. But again, nothing of true importance is shown about the more subtle significance of absence; the question concerning its significance for and among whom, besides humans, is left in the dark. The role of humans is taken for granted; the role of nonhumans is not mentioned, let alone investigated.

Toward the end of his book, Deacon laments the success – intellectually, but also politically and morally – of "a vision of reality that has no place for subjectivity or value" (2012: 544). We are given to understand that Deacon's aim has been to correct this sorry state of affairs. His plea is that we "rethink the frame of the natural sciences in a way that has the metaphysical sophistication to integrate the realm of absential phenomena as we experience them". Speaking of the universe, Deacon notes that "there is more here than stuff. There is how this stuff is organized and related to other stuff. And there is more than what is actual. There is what could be, what should be, what can't be, what is possible, and what is impossible" (ibid.). He goes on to add, "if quantum physicists can learn to become comfortable with the material causal consequences of the superposition of alternate, as-yet-unrealized states of matter, it shouldn't be too great a leap to begin to get comfortable with the superposition of the present and the absent in our functions, meanings, experiences, and values" (2012: 544f.). Yet – and this is the note of melancholy on which Deacon's book ends – "despite the power and insights that we have gained" from what he calls "our current blinkered metaphysical worldview", it "has not helped us to feel at home in the universe" (2012: 545).

There is no denying the importance of what Deacon is searching for on the basis of finding it lacking in the dominant worldview. However, in order to truly challenge the latter, the "place for subjectivity and value" that he misses there would have to sought outside of the purely – solely – human realm, being the prison within which "meanings, experiences, and values" have for centuries been locked in Western thought and science. The true task is precisely not as Deacon defines it; it is much more radical. It is *not* to develop a framework with enough "metaphysical integration to integrate the realm of absential phenomena as we experience them" (2012: 544); instead, it is to develop a framework doing justice to the ways in which *others*, besides us humans, experience absence. Because they do.

A passage in Thom van Dooren's *Flight Ways* speaks directly to this point:

> I am captivated and unsettled by a singular image. It is the image of a penguin returning to a burrow, to a breeding place, that is no longer there or has been transformed so dramatically that it is no longer habitable. Of houses and swimming pools lining the shore of a harbor: people pulled to be closer to "nature", but in a way that makes life for penguins and other harbor dwellers impossible. But this story of returning to lost places is not solely about

> penguins or even coastlines. The returning penguin could just as easily be any number of other migratory or nomadic birds, a sea turtle, or even a seal looking for a haul-out spot on the shore. The beach could be a mangrove, a wetland, a tidal flat, or a range of other settings. All over the world, animals are drawn to return faithfully to places that no longer exist.
>
> (van Dooren 2014: 66)

What van Dooren helps provide here is an important corrective to Deacon: insight into "the realm of absential phenomena" as others – for example, penguins – experience them. It is not only that absence is a reality making a difference to nonhuman species; in the cases – and their number is rapidly increasing – where what used to be present is suddenly absent in the sense of being lost, that absence may spell doom: it is a life or death issue. The research van Dooren is drawing on, as well as that he himself carries out, shows that site fidelity is no less a reality among animals than among humans. For instance, little penguins are *philopatric*, a term that literally means "love of one's home" and that in biology describes a process in which an animal returns to its place of birth or hatching to reproduce. To an anthropologist such as van Dooren, this fidelity to site – not just any site, but *that particular one*, being home to the animals in question – currently (meaning relatively soon after scientific research has truly appreciated its importance) takes on a heartbreaking nature due, mainly, to "the widespread loss of breeding sites that has resulted from the relentless densification of the area" (van Dooren 2014: 75), that is to say, an anthropogenic loss affecting animals in ever-increasing numbers and areas (habitats) in the world; animals who, in many cases and places, continue to return to their distinct home, again and again, only to find that it is gone.

Since life is everywhere about connectivity and in peril once the vast cross-species web of connectivities is threatened, the relationships between predator and prey are spatially and temporarily particular as well as intertwined – meaning synchronized. For one animal to learn to kill another involves learning its temporalities and spatialities; crucially, it is a matter of *being in the right place at the right time*. Created and sustained year after year, synchronies in this sense become a sequence through generational time. "Or at least they used to", as Michelle Bastian points out. For the fact is that, "since the last great extinction event, the tangle and weave of embodied time has grown increasingly ornate and precise, but here in the midst of another such event time is becoming threadbare. The forests, the peccaries, the jaguars, the leatherbacks – all are under threat" (Bastian 2017: 162). Consider that leatherbacks draw on "sequences 110 million years long, knowing where to be, and when, in order to create the beneficial synchronies that make futures" (ibid.) – temporally and spatially *particular* synchronies that, notwithstanding their relative stability through millions of years, now disrupt and disconnect for the anthropogenic reasons alluded to above: overfishing, eutrophication, acidification, climate change, translocation of invasive species, and seabed destruction.

To return to Deacon, he simply gets it wrong at the end of his otherwise impressive book: the appropriate quest in the present circumstances is not for a framework with sufficient "metaphysical sophistication to integrate the realm of absential phenomena as we experience them" (2012: 544). Rather, it is for one capable of sharply illuminating the kinds of absential phenomena that we, partly because we take ourselves to be the only agents having experiences at all, *do not experience*, yet that we clearly should, were the sought framework to qualify as an improvement on the one Deacon rightly – but for inadequate and partly wrong reasons – finds lacking.

The issue overlooked by Deacon – absences indubitably real yet not appreciated as such by the only type of appreciators recognized – is poignantly articulated by Vinciane Despret:

> What the world has lost is not what people mourn. What the world has lost, and what truly matters, is a part of what invents and maintains it as world. The world dies from each absence; the world bursts from absence. For the universe, as the great and good philosophers have said, the entire universe thinks and feels itself, and each being matters in the fabric of its sensations. Every sensation of every being of the world is a mode through which it exists. And every sensation of every being of the world causes all the beings in the world to think and feel themselves differently. When a being is no more, the world narrows all of a sudden, and a part of reality collapses. Each time an existence disappears, it is a piece of the universe of sensations that fades away.
>
> (Despret 2017: 219f.)

But does pointing to ongoing trends of absence in the eminently *physical* sense of an emerging lack of synchronicity between predator and prey, that is, between different species of animals who are dependent on each other's presence (availability) at some particular place and time, really constitute a valid critique of Deacon's understanding of absential phenomena? Am I not ascribing those phenomena a much too literal sense, to be clearly distinguished from the subtle mental one that Deacon sets out to examine?

In my view, taking resort to the old-worn distinction between the physical and the mental so as to keep apart two types of absence will not do. The crucial point is this, and it is no less crucial for being not purely philosophical but empirical, and dramatically so: the myriads of nonhuman beings for whom "absential phenomena" are part of their experience, are themselves becoming increasingly absent from the world that we humans share with them, a world becoming poorer by the day for that loss. Their absence will in many cases prove definitive, non-reversible, in that it is a matter of extinction, mainly human-caused such, yet constituting an absence scarcely noticed by the one species driving it and – ironically – priding itself on being more adept at experiencing phenomena of absence than any other creatures. Doubtless, this ill-conceived pride has played, and still plays, a vital part in facilitating and justifying the practices exacerbating that very extinction. Deacon's conception of nature is incomplete in not paying sufficient attention to this dimension of human-caused absences in nature.

XI

Return to Kohn. If the ability to experience absential phenomena is often taken as an exclusively human one, the same goes for representation. Likewise, contends Kohn, we suppose that all representation has language-like properties. He maintains that this conflation of representation with language – the assumption that all representational phenomena have symbolic properties – holds for "even those kinds of projects that are explicitly critical of cultural, symbolic or linguistic approaches". Thus "the alignment between humans, culture, the mind, and representation, on the one hand, and nonhumans, nature, bodies, and matter, on the other, remains stable even in posthuman approaches that seek to dissolve the boundaries that have been erected to construe humans as separate from the rest of the world" (2013: 40). Kohn mentions Bruno Latour and Jane Bennett as two cases in point; I shall come back to both.

What, then, does it mean to radically rethink what it is that we take representation to be? To begin to grasp what it means, says Kohn, we need first to "provincialize language"; namely, in Viveiros de Castro's words, to "decolonize thought", in order to see that thinking is not necessarily circumscribed by language, the symbolic, or the human (de Castro quoted in Kohn 2013: 41). Only then will we be able to take a fresh look at who in this world "represents", as well as what it is that counts as representation. We need to understand how different kinds of representation, involving different agents, interact with each other, allowing us to explore what signs look like beyond the human.

Thus the point is reached where Kohn starts to address the notion of agency "beyond the human" that is at the core of my suggestion that animism is panpsychism in practice. Kohn's general statement is worth quoting:

> We humans are not the only ones who do things for the sake of a future by representing it in the present. All living selves do this is some way or another. Representation, purpose, and future are in the world – and not just in that part of the world that we delimit as human mind. This is why it is appropriate to say that there is agency in the living world that extends beyond the human. And yet reducing agency to cause and effect – to "affect" – side steps the fact that it is human and nonhuman ways of "thinking" that confer agency.
>
> (Kohn 2013: 42)

Kohn follows Peirce and his notion of "thirdness" in holding that the generality of thought is that it is not just located – imprisoned within – a single stable self. Peirce puts it like this: "Man is not whole as long as he is single [;] . . . he is essentially a possible member of society. Especially, one man's experience is nothing, if he stands alone. If he sees what others cannot, we call it hallucination. It is not 'my' experience, but 'our' experience that has to be thought of; and this 'us' has indefinite possibilities" (Peirce CP 5.402). This "us" is general, explains Kohn, in keeping with his assertion cited earlier, that there is not one "we" but many.

The many we's inhabit the tropical forest and help make tropical life what it is. That life, in its entirety, is made up of a richness of proliferating webs of habits. Such habits, Kohn submits, are real, "regardless of whether or not I can appreciate them". It is by way of acquiring, through human-nonhuman encounters such as experienced in, for example, hunting, that, say, an individual agent – be it human or nonhuman – can potentially become aligned with a broader "us". In a passage that eloquently blends elements of animism and Peirce's semiotic philosophy, where the concept of habit originates, Kohn writes:

> Like our thoughts and minds, birds and plants are emergent reals. Life-forms, as they represent and amplify the habits of the world, create new habits, and their interactions with other organisms create even more habits. Life, then, proliferates habits. Tropical forests, with their high biomass, unparalleled species diversity, and intricate coevolutionary interactions, exhibit this tendency toward habit taking to an unusual degree. For people like the Avila Runa, who are intimately involved with the forest through hunting and other subsistence activities, being able to predict these habits is of the utmost importance.
>
> (Kohn 2013: 62)

On this view, then, to be alive and to continue being part of the life of everything living, involves each living creature's aligning itself with an ever-increasing array of emerging habits. The life-sustaining role of habit is, however, not exhausted in the form of reliance on regularities, on repetition allowing for prediction – when A, then B – since life also to a large and never-to-be eliminated extent consists of disruption and shock. To help theorize this, Peirce proposes that as opposed to inanimate matter, designated as "mind whose habits have become fixed so as to lose the powers of forming them and losing them", mind (or self) "has acquired in a remarkable degree a habit of taking and laying aside habits" (CP 6.101; see Kohn 2013: 62). Kohn gives the example of a Runa Hunter's feeling of bewilderment when the supposedly dead peccary suddenly jumps up and runs off. In keeping with a well-known insight in phenomenological philosophy, it typically takes a situation of "breakdown", of the absence of that whose presence, and functioning, we take for granted, to become aware of it and its crucial significance for our dealings in the world. Kohn's Peircean way of expressing this is to say that it is in such moments of "shock" – e.g., the "dead" peccary jumping – that the habits of the world make themselves manifest. The point is that "we don't usually notice the habits that we in-habit". It is only when the world's habits clash with our expectations that the world in its otherness is revealed. The challenge that follows this disruption is to grow; "to create a new habit that will encompass this foreign habit and, in the process, to remake ourselves, however momentarily, anew, as one with the world around us" (Kohn 2013: 63). It should go without saying by now that such remaking holds not only for humans but for all those life forms that partake in the webs of habits involved.

Against this background, we can appreciate Kohn's take on animism: "If thoughts are alive and if that which lives thinks, then perhaps the living world is enchanted. What I mean is that the world beyond the human is not a meaningless one made meaningful by humans" (72). Contrary to the conventional view that the world beyond humans is devoid of meaning and passively "awaits" the ascriptions and projections of such that presumably only humans can author, "mean-ings – means-ends relations, strivings, purposes, telos, intentions, functions, and significance – emerge in a world of living thoughts beyond the human in ways that are not fully exhausted by our all-too-human attempts to define and control these" (ibid.). Kohn proceeds to make the claim that "the forests around Avila are animate", explaining that "these forests house other emergent loci of mean-ings, ones that do not necessarily revolve around, or originate from, humans. This is what I'm getting at when I say that forests think" (ibid.). Here, Kohn reconnects with his overarching idea that "if thoughts exist beyond the human, then we humans are not the only selves in this world", which is to say that we are not the only kinds of *we*. "Animism, the attribution of enchantment to these other than-than-human loci, is more than a belief, an embodied practice, or a foil for our critiques of Western mechanistic representations of nature". Thus we should not just ask "how some humans come to represent other beings or entities as animate; we also need to consider more broadly what it is about these that make them animate" (72f.).

I am critical of Kohn's formulation that animism is "the attribution of enchantment to these other-than-human loci", since human subjects, and they alone, are so easily taken as the only ones in the habit of – and so capable of – performing such attribution. Kohn's opposition to the anthropocentrism of this notion would in my opinion be better served by talking about meaning as forthcoming from, as well as interpreted by, and responded to, by a multitude of nonhuman selves. And whereas "attribution" buys into the wrong epistemological model, "enchantment", with its strong Weberian baggage, likewise suggests something that only humans are engaged in. We have learned from Ingold that animism, for those living their lives in tune with it, is not first and foremost about representation – understood as a certain way of depicting the world – but about doing justice to one's peculiar, species-dependent role and place within a world that in its entirety is saturated with powers of agency, each one participating in so many species-specific relationships, and *experienced and acted upon as such*, to paraphrase Ingold (2011a: 14). In putting its essence like this, I do not mean to oppose Kohn's understanding of animism; I only want to express it in a way that avoids anthropocentric language and highlights its qualities as a *lived* reality. In saying this, I have not forgotten that Kohn is eager to free representation both from its linguistic colonization and from its exclusively inter-human realm.

To illustrate his point about other-than-human loci of meanings, Kohn invokes not a monkey in its encounter with a human hunter, but a dog. The example is that of Runa, of women who "certainly felt that they were able to interpret the dog's barks, but that's not what makes them recognize their dogs as selves. What makes their dogs selves is that their barks were manifestations of their interpretations of the world around them" (2013: 73). The women are very aware that how these dogs interpret the world around them, matters vitally. Clearly, we humans are not the only beings who actively interpret the world and who author actions as well as responses as part of that ongoing interpretation – ongoing since its cessation would immediately jeopardize one's existence in that world that is inhabited by/with so many others, all bent on pursuing *their* goals.

Kohn's take on this is again strikingly Peircean: "'Aboutness' – representation, intention, and purpose in their most basic forms – is an intrinsic structuring feature of living dynamics in the biological world. Life is inherently semiotic" (2013: 74). Life, that is to say, is a sign process. Kohn cites Peirce to argue that any dynamic in which "something . . . stands to somebody, for something in some respect or capacity", as Peirce definition of sign determines, would be alive (CP 2.228; see Kohn 2013: 74). Kohn stresses how Peircean semiotics moves beyond the conventional understanding of signs as something that "stands for something else", highlighting that signs stand for something in relation to a "somebody". And significantly for Kohn the anthropologist studying the Runa cosmology and way of life (two sides of the same coin), this "somebody" – or self, as he prefers to call it – "is not necessarily human, and it need not involve symbolic reference, subjectivity, the sense of interiority, consciousness, or the awareness we often associate with representation for it to count as such" (75). It follows that, on Kohn's view, selfhood is not limited just to animals with brains; "plants are also selves". Nor is selfhood coterminous with a physically bounded organism; selfhood, that is, "can be distributed over bodies (a seminar, a crowd, or an ant colony can act as a self), or it can be one of many other selves within a body (individual cells have a kind of minimal selfhood" (ibid.).

Note the implication that there can be a "self" without that self being, or having, subjectivity. Kohn declares being in agreement with Deacon about this; Deacon who writes that "self is what defines the very individualization that characterizes an organism and is its most fundamental organizing principle. Only living organisms are truly individual in the sense that all aspects of their constitution are organized around the maintenance and perpetuation of this form of organization". Organism functions, then, are indirectly self-referential and self-projecting. By contrast, continues Deacon, "in the inanimate world around us, we find no trace of this circularity of generative processes. Though bounded and unified, neither stones, not drops of water, nor automobiles, nor computers, nor any other non-living artifact is reflexively individuated in this way" (Deacon 2012: 465). Subjectivity, for its part, is "almost certainly a specially developed mode of self that is probably limited to creatures with complex brains" (2012: 466). The upshot is that whereas "even organisms as simple as bacteria have properties that qualify them as selves, in at least a minimal sense . . . subjectivity is not a critical defining property of self" (ibid.).

Unique and central to life, to all its instantiations, is a play of "remembering and forgetting"; according to Kohn, any lineage of living organism – plant or animal – will exhibit the back and forth between the two. And whereas for instance anteater lineages selectively remember their previous fits to their environments, Deacon's stone or drops of water do not; nor do snowflakes. What is crucial is that a self is able to *act for itself*; most fundamentally, it does so in that it "strives to maintain its form" (Kohn 2013: 76).

To understand what these general statements about self, subjectivity, and agency mean with regard to my suggestion that animism is panpsychism in practice, we need to revisit once again the eminently concrete phenomenon with which Kohn's study started: the practice of hunting.

XII

Kohn begins his book with an anecdote about the jaguar. About to join the Runa for a hunting trip, he was told to remember to sleep face up. Why? Well, "if a jaguar were to pass by he would see me as being capable of looking back and would leave me alone". Were he to forget this advice and sleep facedown, the cost might be high: a potential passing jaguar would probably treat him as prey and attack him. Thus is brought out, in terms of life and death, the fundamental fact that *how an animal like the jaguar sees us matters to us, since it matters to them*, determining what they will do: walk away or attack. Positioning ourselves, as human subjects, in such a manner as to communicate to the animal that the animal is another subject invites it – say, the jaguar – to treat one as a predator like itself. Negatively put, one survives by making sure not to be noticed, identified, or responded to as prey.

But, the reader may well ask, have we not said what needs to be said about this already? True enough, I mentioned the importance of being able to turn certain others into "non-selves", indeed the imperative of doing so simply to be able to survive. However, drawing on the material we have looked at since then, we are now in a position to appreciate the issues involved in a more complex way.

We know that the Runa, like other Amazonians, treat jaguars, peccaries, monkeys and many other nonhuman beings with whom they co-inhabit the forest as soul-possessing, signifying, intentional selves. In doing so, the Runa are animists; for them, nonhumans are animate, meaning they are persons and must be communicated with and treated accordingly. Denying them person-status, as the anecdote amply illustrates, would only be at the denier's peril, echoing a point made several times already: that denial tends to consume its author and prove self-destructive. Even if such denial may well be a possibility in principle, in real life it is no option. The Runa, it might be said, in a Peircean vein not used (for once) by Kohn, are in the individual and collective habit – amounting to a life form – of taking the person-status of the nonhuman creatures that they encounter for granted.

Kohn credits Descola and Viveiros de Castro with enriching anthropology's understanding of animism in that they not only ask "how other humans come to treat nonhumans as animate", as did and still do many of those – counting philosophers – who merely use animism as a foil for critiquing Western mechanistic representations of nature (though not mentioned by Kohn, Horkheimer, and Adorno's *Dialectic of Enlightenment* may serve as a case in point). The role played by the jaguar in the anecdote highlights what is wrong, or in any case fundamentally incomplete, in such an understanding of what animism is, or better, does. If animals such as jaguars, or monkeys, *also represent us* and as part of doing so anticipate, say, the hunter's next move, his intention in bringing down the palm tree, and so respond to what Deacon terms "absential phenomena", fleeing the scene while it is still time, which is to say *before* the hunter's next move is fully out in the open, while it still has the quality of a potentiality, *a not-yet* and as such something that is not rather than something that is, negativity not positivity, "then we cannot just ask how it is that some of us humans happen to represent them [as representing us]" (2013: 94). Runa animism is grounded in what is an ontological fact as fundamental and as ineluctably given as any: "that there exist other kinds of thinking selves beyond the human". That is why, says Kohn, he calls his book "How Forests Think", not "How Natives Think about Forests": "if we limit our thinking to thinking through how other people think, we will always end up circumscribing ontology by epistemology" (ibid.). It suits my purposes that he makes this point and puts it precisely in these terms, since it echoes what I, drawing on Charles Taylor, pointed out in my discussion of Nagel in Chapter 1: the tendency in philosophy since Descartes' *cogito ergo sum* to privilege what we can know over what is – or alternatively, adjusted to the present context, to privilege what we know (or don't know) about what other-than-humans know (or don't know), over what they are.

In claiming that living beings are loci of selfhood, Kohn wants to make a general claim about selfhood, and he says that he makes the claim "empirically" in that it "grows out of my attention to Runa relations with nonhuman beings as these reveal themselves ethnographically. These relations amplify certain properties of the world, and this amplification can infect and affect our thinking about the world" (2013: 94). He goes on to comment that whereas "for us" – presumably Westerns – the machine is the model of the universe, for animists the animal person is. Which view is the true one? Kohn's answer is that "ontologically speaking, each has its own truth: animals are persons, and there are things about the world that do resemble partible machines (which is a reason why reductionist science is so successful)" (ibid.). However, Kohn is not concerned to distinguish between true and false with respect to what is commonly regarded as two competing, and possibly incompatible, world views – partly, I take it, because the models of mechanism (or materialism, or naturalism, or physicalism) on the one hand and that of animism on the other are not first and foremost *views* in some theoretical or purely mental sense. Rather, and in keeping with my overall point, they exemplify two specific kinds of what Kohn refers to as "engagements", eminently practical, lived, acted-upon such in both cases, "based on certain presuppositions that themselves grow out of those engagements, amplify unexpected and real properties of the world that we can harness to think beyond the human as we know it" (2013: 95). As should be clear by now, Kohn's use of the verb "think" and the noun "thought" emphatically does not conform to the model of representation still reigning in Western philosophy: those doing the thinking and thereby contributing from their respective viewpoints to the way forests think, do so in ways embedded with and inseparable from their doing, that is, from their interactions with others in all sorts of different forms, shapes, and guises. In order to live as a group of humans amid a manifold on nonhumans in the forest, the Runa engage intimately with all the beings in the forest, finding ways to "enter this vast ecology of selves to harness some of its plenty" (95). To do so successfully – ultimately, to survive – the Runa need to recognize the importance of their continuities with, as well as differences from, all those particular others. Differences are never absolute or complete, never the whole story.

As lived, then, animism is not merely or even primarily about how humans come to see nonhumans as persons, as loci of selfhood. To focus on this, to take it as the essence of animism, is to seize on one side of the coin as it were and miss out on the other. It reveals a "Western" take on animism, seeing it through the optics of what is "exotic". It is the othering of animism. Though animism is about that too, i.e., about humans seeing nonhumans as persons, it is first and foremost about recognizing – in practice – that, as a matter of fact, not theory, all sentient beings, be they spirit, animal, or human, see themselves as persons and enact this fact about what they *are* in everything that they *do*. So, to take the example of hunting, animals, not only humans, go hunting. And to do so successfully, they engage in various "tricks of the trade", relying not only on instinct but to a large extent on *in situ* learning by doing (Deweyan style), practicing to perfection the tricks that best match their species-specific skills and capacities as hunter-selves. As Kohn reports, in another anecdote from the Runa, "one man took delight in explaining to me how the giant anteater adopts the perspective of ants in order to fool them; when the anteater sticks its tongue into ant nests, the ants see it as a branch and, unsuspecting, climb on" (2013: 96). All across the spectrum-continuum of living creatures, the myriad loci of selves, there is the never-pausing, never-ending need to capture – anticipate and accordingly respond and adjust to – the perspective of the other selves, to treat all alike as vital parts of the larger whole. Attentiveness to the points of view of other organisms is an imperative of existence, a given inseparable from the givenness of all those others with whom each one in particular shares its existence in the forest, on Earth.

I already commented on the perils of refusing to recognize the other(s) as person(s). In the particular ecology of selves that inhabit the cosmos that is the forest, selves must recognize what Kohn terms the "soul-stuff" of other selves in order to interact with them. Echoing Hegel's famous theory about the intersubjective preconditions of recognition, of there coming to exist such an entity as the (human) self, Kohn observes that "to remain selves, all selves must recognize the sound-stuff of the other souled selves that inhabit the cosmos" (2013: 117). As mentioned earlier, he employs the term *soul blindness* to describe "the various debilitating forms of soul loss that result in an inability to be aware of and relate to other soul-possessing selves in this ecology of selves". Since in this ecology all selves have souls, "soul blindness is not just a human problem; it is a cosmic one" (ibid). That "our lives depend on our abilities to believe in and act on the provisional guesses we make about the motivations of other [nonhuman] selves" (118), is a fact of life unquestioningly accepted among the Runa.

It says a lot about "our" (modern, Western, anthropocentric) cosmology that we tend to regard the acceptance of that fact as evidence of a "primitive" culture that has yet to become "modern" and scientific in Weber's sense, or – alternatively – as something exotic to which one is attracted, a long-lost life-form for which one feels nostalgia, casting its nowadays often marginalized practitioners in sentimental terms and fearing – to a large degree correctly – that they embody a cosmology and life-form on the brink of extinction. This exemplifies the intimate yet increasingly precarious connection between *loss of biodiversity and loss of cultural diversity* in the era of global capitalism (see Vetlesen 2015). The connection is such that there is no way to preserve, say, an ecologically intact and biologically rich forest without preserving at the same time and as part thereof the people who perceive and treat that forest as their one and only home; to destroy the one is to destroy the other.

That said, there is something odd about how Kohn, the currently Montreal-based anthropologist doing ethnographic fieldwork in Ecuadorian Amazonas and reporting his experiences and findings to a well-educated (predominantly) Western audience, spends page after page making his case for the claim that dogs bark in order interpret the (their) world, or that a monkey is able to anticipate the human hunter's next move so as to evade it, or that a jaguar distinguishes between potential prey according to whether they (it) appear like selves or like objects. Every child in the woods knows this and has done so as long as there have been humans in the woods. The *agency of all sorts of creatures*, I submit, is the most fascinating, self-explanatory, self-manifesting thing (double meaning intended), and phenomenally and experientially the most indubitable fact in the world. Yet it is precisely the one that Descartes made a career (indeed a philosophical revolution) of declaring dubitable *par excellence* (probably to the absolute disbelief of the non-Westerners whose opinion were not sought); a given since time immemorial, as shown by Paul Shepard in his *Nature and Madness*, long preceding the advent of humankind on Earth, giving rise to wonder, awe, and the playing out of curiosity ever since: what will that animal do if I . . .?

Kohn's meticulous renderings of his experiences with the Runa betray that he is cognizant of the way in which the burden of argument has shifted in the course of the historic shift from "mythical" to "modern" cosmologies Weberian style. He knows, that is, that his book's audience will overwhelmingly consist of people whose lives, relationship to, and views about nature and nonhuman animals will contrast sharply with those found among the Runa. It is vital to recognize that this goes both ways: insofar as Runa cosmology strikes us as exotic, our cosmology – in which agency and selfhood are denied to nonhumans – will strike the Runa as no less so (or perhaps: as even more so). But apart from on a purely abstract, philosophical level, the symmetry suggested by the mutuality of qualifying as "exotic" is in fact all about asymmetry: time is running out for the nonmodern cultures still existing to provide an anthropologist like Kohn with his research material: a culturally and biologically intact ecosystem such as the tropical forest inhabited by the Runa, a forest rich enough in biodiversity to demonstrate the thesis that forests think, and the testimony to cultural diversity – cosmological richness – that a culture practicing the notion that forests think embodies.

Given the threat of extinction to which a life form such as that of the Runa is subject in today's world, it may come as a surprise that Kohn does not pay more attention to it. And yet he does – only subtly. His book raises fundamental philosophical questions, more fundamental and more philosophical than most of what passes for philosophy these days. As far as I can tell from his analysis of Runa cosmology, Kohn is a realist of sorts, though the precise sort in question is hard to spell out in detail. Kohn is acutely aware that ours is culture of constructivism, academically even more so than commonsensically (to recall Griffin in Chapter 2). We are socialized into thinking that as human agents endowed with the gifts of selfhood and intentionality that nonhumans (supposedly) lack, we cannot help but project our notions about the world outside us onto that world and its various entities, lending everything we think and talk about a *for-us* dimension that cannot be avoided and that will therefore, in principle, always and everywhere cause the *in-itself* dimension of all entities not-human to be beyond our reach – not literally, but epistemologically and therefore (as the hegemonic logic of priority has it) ontologically. Is there mind in matter? Interiority in trees? Does a river remember? To all these questions, the answer is and remains that given by Russell all those years ago: we will never know.

In his short but important Epilogue, Kohn, in his pointedly non-Russellian way, addresses the questions just mentioned. "What is this world beyond us and the sociocultural worlds we construct?", he asks (2013: 223). He does not give a direct answer. Instead, he says that he has sought to offer a "reflection on how we might think about reality as something that extends beyond the two kinds of real that our dualistic metaphysics provides us: our distinctively human socioculturally constructed realities, on the one hand, and the objective 'stuff' that exists out there, beyond us, on the other". He goes on to state, somewhat cryptically, that he has deliberately placed the realm of society and culture "on the first, and hence the right, hand" and relegated "the realm of things to the second hand", adding that "it is that which we take to be human (our souls, our minds, our cultures) that currently dominates our dualistic thinking". As a result, "the realm of others, the nonhumans (evacuated of animacy, agency, or enchantment)" is consigned to the left hand, a hand that nonetheless, Kohn comments, has "its own subversive possibilities" (ibid.).

Interestingly, Kohn does not portray this dualism as "just a sociocultural product of a particular time or place", as so many critiques of Western metaphysics typically do, taking themselves to be subversive to the extent that the dualism that is held to be "universal" is shown, by way of deconstruction, ideology critique, and the like, to be nothing of the sort but highly selective and contingent instead. Implicitly dismissing this all-too conventional trope, Kohn contends that "our propensity for dualism (our 'twoness', in the Sphinx's terms) is the product of the distinctive properties of human symbolic thought"; hence it is that thinking "creates systems of signs that can come to seem radically separate from their worldly referents" (223). Thinking in twos, then, is "ingrained in what it means to be human" (ibid.). But Kohn does not want to leave it at that. Dualism in thinking may be deeply human and so not a tendency, let alone ideology, for which, say, Descartes can be held responsible; rather, it goes back as long as we thinking humans do, if I understand Kohn correctly (he does not say). Moving beyond what he describes as "this kind of handedness" *is* possible, it is something that we can do, indeed *ought* to do, while there is still time. Yet, it "requires a real feat of defamiliarizing the human", namely, to "undertake an arduous process of decolonizing our thinking", to "provincialize" language in order to "make room for another kind of thought", one that is "more capacious, one that holds and sustains the human". What kind of thinking is he referring to? We know the answer: it is the one that forests do, "the kind of thinking that thinks its way through the lives of people, like the Runa (and others), who engage intimately with the forest's living beings in ways that amplify life's distinctive logics" (224).

Leaving constructivism and the hegemony commonly claimed for it aside, then, the realism I spot in Kohn is one captured in his assertion – articulated matter-of-factly rather than as a bold claim inviting the whole onus of proof – that "the multispecies ecology that I have described here is real in an ethnographic and ontological sense" (227). But do not get his realism wrong. The realness – factuality – of that ecology, a very particular one in time and space, is not *sub specie aeternitatis*: it is not a given that, qua real, is bound to remain such. Qua reality, earthly such, it is – in my words, not Kohn's – an eminently precarious, endangered piece of reality, enjoying the ontological condition of reality here and now, yet in no way guaranteed to continue doing so in the future. In Kohn's again suitable matter-of-factly way of putting it, the multispecies ecology of selves he has reported from/on in the case of the Runa is one that "depends for its existence on the continuous flourishing of dense nonhuman ecologies just as it does on the humans who live by tapping those ecologies" (237). The implication is unequivocal: "If too many of these elements that make up this ecology of selves disappear, a particular kind of life (and afterlife) will come to an end – forever. And we will have to find ways to mourn its absence" (ibid.) – recalling the kind of mourning I found absent in Deacon's framework, a mourning involving so many beings and life-forms beside our human one, always one among many.

It may sound banal, yet it is fundamental: everything to be learned by and from "how forests think" will be lost when the forests are lost. It is a thinking that is inseparable from its element; a mindset inseparable from its physical comportment. We may writ large what the environmental philosopher typically will say about tigers: if you want to save the tigers, you have to save the forests; you cannot have one without the other. What Kohn the anthropologist may teach the philosophers debating panpsychism, then, is this: the question about what or who exhibits mind, or mentality, or interiority, such as to posit a specific point of view on/in the world, is neither to be raised nor settled in the abstract, but on the ground, bottom-up, not top-down. It is an utterly concrete question; it is so because it needs to be asked with particular entities – existents – in mind (sic), always and everywhere. Those who exhibit the capacities that correspond to *the one side* of the dualism Kohn described, depend on *the other side* for their continuing to exhibit them, alternatively referred to by Kohn as "the objective stuff that exists there beyond us" and as "the realm of the others, the nonhumans", as we recall.

And yet, there is a sense in which Kohn, deliberately or not, challenges this way of looking at the issue of what is real and what it takes to remain so. It is as if he does not want it to be the – his – last word. Consider his remark, bordering on the enigmatic, that "the world is animate, whether or not we are animists" (217). What does he mean? The world, Kohn goes on to elaborate, "is filled with selves – I daresay souls – human and otherwise. And it is not just located in the here and now, or in the past, but in a being in future – a potential living future" (217). When he contends that "the realm of the masters is real", and that "spirits are real", Kohn wants to drive home his thesis that "reality is more than that which exists" (216).

Even though he does not mention it at this point, I assume that Deacon's notion of absential phenomena – generally: absence, what is not, or not yet – is one of the things on Kohn's mind, allowing him to endorse a conception of reality that covers more than the totality of entities existing, or states of affairs being either true or false, at one particular time. There is more to reality than captured by, or reducible to, factual presence. In saying that spirits are real, Kohn warns against "taking spirits to be a kind of real – the kind that is socially or culturally constructed – that is 'all too human' and all too familiar". To illustrate, he accepts that "gods emerge with human practice", but hastens to add "that does not make them reducible to or circumscribed by the human contexts in which such practices unfold" (216). The upshot seems to be that, while human practices, as part of the multispecies ecology of the forest, have a crucial role to play in sustaining the way in which gods, or spirits, or the selfhood of nonhuman animals or plants, are real – real in my Taylor-inspired sense of "that which won't go away" and of demanding to be taken into account, to count as so many points of view in/on the world, making it what it is – *these entities are also real in a human-independent sense*, in keeping with realist ontology. Not of our making, we may nevertheless break them, undo them, in that we have the power – in combination with the technology and intelligence (or lack of it, in the form of lack of wisdom) – to physically destroy the forest that is their home – just as we destroy the homes that van Dooren's penguins try, in vain, yet again and again to return to.

5 Agency posthumanist style

Proliferation or decimation?

Discussing Michel Serres' *The Natural Contract*, I quoted his assertion that today "we are disturbing the Earth and making it quake", commenting that "now it has a subject once again" (1995: 86). No longer "fixed and immemorial", the Earth is "trembling". It is so, Serres tells us, "because it is being transformed by our doing" (ibid.).

But the subject as well as the doing Serres is referring to remain conspicuously unspecified. He offers suggestive ideas and sweeping claims, not an accurate analysis of how the transformation of the Earth he is talking about has come about and what particular forces and agents are causing it. A "bond" to nature that used to be highly local has been lost in our modern technological society, but apart from that we are left in the dark about what led to the disturbance that now makes the Earth quake. What is clear, though, is highly important: we are causing disturbances of such a magnitude as to make the Earth *respond.*

In this chapter, I explore the wider implications of this phenomenon. In doing so I engage with what is today probably the most influential proponent of posthumanist thinking, Bruno Latour. Just like this book as a whole, this chapter moves from the theoretical to the practical, addressing the much-debated understanding of agency in an increasingly empirical – historical and political – context. I shall employ insights from the anthropological studies introduced above with a view to showing their relevance as correctives to central tenets within posthumanism.

I

In the second of his eight lectures on the "new climate regime" making up his book *Facing Gaia*, Bruno Latour seizes on Serres' reformulation of Galileo's famous line "*Eppur si muove!*", translated as "and yet it moves". In what Serres depicts as Galileo's "second trial", the new and, according to Latour, worrying twist is that the Earth is moved instead of moving, prompting as we saw Serres' remark that "it has a subject once again". Latour observes that what is shocking today is "to learn that the Earth has become – has become again! – an active, local, limited, sensitive, fragile, trembling, and easily irritated envelope", a fact ridiculed or outright denied by "the new Inquisition (henceforth economic rather than religious)" bent on "making fun of this new as well as very old animated and fragile Earth" (Latour 2017: 60). As climatologists and other experts are telling us, whereas Galileo's Earth could revolve, it had no "tipping point", no "planetary frontiers", no "critical zones". In short, Latour submits, "it had a *movement*, but not a *behavior*"; hence "it was not yet the Earth of the Anthropocene" (ibid.).

What is truly new, then, is that the Earth exhibits "behavior". This fact compels us to take a fresh look at the notion of agency – not first and foremost *human* agency, as the term for the new geohistorical era – the Anthropocene – might suggest, but agency in all its different forms and manifestations, whereby the nonhuman such will prove potent in ways never witnessed before. For my purposes, the renewed interest in nonhuman forms of agency sparked by the entry into the Anthropocene helps illustrate, in a dramatic fashion, my overall argument about the relevance of animism and panpsychism, being cosmologies with a keen philosophical as well as lived appreciation of agency in all its different instantiations.

What Latour admires in Serres is that he directs our attention toward "the astonishing *connivance* between formerly distinct agencies – as opposed to one another, as were the old figures of object and subject – that are now so mixed" (2017: 61). And yet, what I find lacking in Serres' book is precisely an informative account of the distinct agencies in question. Can it be that Latour has noticed something that I am missing?

In the present era, says Latour, the traces of our action are visible everywhere. This is, on the face of it, a pretty common sense view of what the Anthropocene is all about: a series of geophysical changes on planet Earth, *to* planet Earth, caused by human-initiated activities, whether knowingly and intentionally or not. This being so, what follows for agency and the way it now must be understood, as distinct from in the Holocene and other previous eras?

Throughout his lecture "How not to (de-)animate nature", Latour circles around this question, yet without really answering it. Consider the following passage: Today, "just as happens in prescientific and nonmodern myths, we encounter an agent that takes its label, 'subject', from the fact that it can be *subjected* to the whims, the bad moods, the emotions, the reactions, and even the revenge of another agent, which also takes its quality as 'subject' *from the fact that it is equally subjected to the action of the other*" (2017: 62). He immediately adds: "Being a subject does not mean acting in an autonomous fashion in relation to an objective context; rather, it means *sharing* agency with other subjects that have also lost their autonomy" (ibid.).

This is a remarkable passage. For a start, it appears less than consistent to go on using, as a key notion, the term "subject" just after asserting that "the old figures of object and subject" will no longer do. Second, in referring (in a footnote) to Eduardo Kohn's *How Forests Think* when making the point that what we witness today is "just as happens in prescientific and nonmodern myths", namely that "an agent takes its quality as a 'subject' from the fact that it is equally subjected to the action of the other", one wonders how Latour arrives at such a one-to-one comparison-cum-equalization between the animism lived by the Ecuadorian Runa on the one hand and how agency – in a broad sense – figures and is reconfigured in the Anthropocene, on the other. Third, the suggestion that being a subject means "sharing agency with other subjects that have also lost their autonomy" does seem to capture an important insight going back (at least) to Hegel's theory about recognition and its peculiar logic of reciprocity: to be real and effect what it needs to effect, recognition must be mutual, never a one-sided affair. Even so, there is an ambiguity in talking about "subjects that have also lost their autonomy". Strictly speaking, in order for a subject to lose autonomy, it must at some point have possessed it. But that is precisely the notion that Hegel – and scores of critics of atomism, individualism, and mechanism – debunks, exposing it as an illusion (Hegel 1977, 1991; Williams 1997). Perhaps what Latour is addressing is something else, namely that subjects come to lose the *belief* that they are autonomous, realizing its falsity. Regrettably, he does not explain what he has in mind. Nor does he offer a definition of autonomy. For lack of such, the connotations of separability, self-sufficiency, and independence that the notion of autonomy conventionally carries, invite said objections.

Let us see if Latour becomes clearer as he unfolds his argument. He tells us that "as soon as we come close to nonhuman beings, we do not find in them the inertia that would allow us, by contrast, to take ourselves to be agents, but, on the contrary, we find agencies that are *no longer without connection* to what we are and what we do" (2017: 62). Unfortunately, the "no longer" in this formulation is another case of ambiguity: were the agencies in question really ever without such connection to us; or is it rather that the belief – held by whom, exactly? – that they were so now is exposed as false? As for the former: was there ever a situation, a time or a place where human and nonhuman agencies coexisted, yet were not connected? If anything, ethnographic studies such as Kohn's go to show that the interconnectedness between different sorts of agencies – in terms of how they interact and mutually shape each other and depend on each other, yet how they may also be in conflict – is inseparable from the entire history of human presence on planet Earth. It captures a fact, a condition of life, fully acknowledged in pre-modern and nonmodern cosmologies, yet disputed in specifically modern ones at least since the intellectual revolution brought about by Descartes.

Pointing this out may appear pedantic in a manner totally in mismatch with Latour's style of thinking and writing. Faulty logic and lack of conceptual rigor aside, he is surely making a valid point: the connection between agencies human and nonhuman is becoming much more pronounced, as well as more unpredictable and dangerous, than at any earlier point in history (see Merchant 2016). That is the important change that Latour wants to direct our attention to.

The circumstance that human action is now visible everywhere, that nothing appears to be outside its reach and unaffected by it, does not sit well with the way we "moderns", subscribing to a scientific as opposed to "mythic" world view (to invoke Weber's distinction once again), typically approach things: whereas it now becomes abundantly clear that the Earth, as Latour says, is "animated by countless forms of agents, we have sought to conceptualize it as essentially inert and inanimate" (2017: 63). That we have done so for the last four centuries is something there is no need to detail here; suffice it to recall the discussion of Whitehead's account in his *Science and the Modern World* in Chapter 2. In Latour's take on this, however, natural law plays a part not addressed by Whitehead. It does so due to Serres, considered by Latour to have exposed the weakness of natural law, which consists in saying that "there is indeed law in nature" in a prescriptive sense, and nevertheless "law, true law, is found only on the other side, in culture" (2017: 63). This prompts Serres to ask: "What language do the things of the world speak, that we might come to an understanding with them, contractually?" (Serres 1995: 39). Considering that the old social contract, the one exclusively between humans, was in fact "unspoken and unwritten", the contrast with the "order of nature" is not at all as great as normally assumed, for "in fact, the Earth speaks to us in terms of forces, bonds, and interactions, and that's enough to make a contract" (ibid.).

It is tempting to comment that the asserted fact that "the Earth speaks to us" in such a manner is something to which we have been studiously deaf and blind, proudly so in taking the muteness of nature to be part and parcel of its alleged "intrinsic" inanimate essence, denouncing as unenlightened or insane those who would defy that notion and insist that nature "speaks" and that they can hear the pain expressed. In other words, as long as the Earth's capacity to speak to us in Serres' sense is being denied, the basis on which to make a contract is simply lacking. So, if we are nowadays prepared to grant that the Earth does speak to us, such preparedness is proof of a change on our part: for so long, we did not listen, and did not hear anything; now we do.

Though correct, putting it like this is to miss the point – the point that both Serres and Latour want to make. The change that really matters here, and with reference to which everything of true importance will be decided in the years to come, is not strictly speaking a change taking place in us but in the nature of which we are inextricably part but that nevertheless consists of myriads of agents different from us, from our peculiar capabilities of agency. To put it in terms inspired by James Lovelock, the father of Gaia theory: the forces, bonds, and interactions in terms of which the Earth presently, in the Anthropocene, speaks to us are not, qualitatively, of *the same* sort and order as they used to be until recently. Rather, the Earth – or Earth system, as it is now scientifically termed – is *itself* changing. It had a movement, and now it exhibits behavior, a behavior let loose by human-initiated activities yet increasingly developing a momentum all its on own, so that, for all its being caused (predominantly) by human agents, the momentum is not, never was, and probably (geoengineering fantasies notwithstanding) never will be controlled by humans.

To put it in my own words rather than Serres' and Latour's, this then is the new dialectic playing itself out: the result of humans' treating the Earth Cartesian-like as matter, and so as inanimate, inert, and – essentially – unchangeable, now turns out to have produced an effect for which the cosmology accompanying, and legitimizing, that thing-like, means-oriented treatment is eminently ill-prepared: the effect being the animation – or should we say reanimation? – of all sorts of forces denied agency. Having been denied, and thereupon mistreated, exploited, subjugated for so long, those nonhuman agencies now strike back at us, doing so with such vehemence as to prompt scholars (such as Lovelock and, though less decidedly and unequivocally, Latour) to speak of their "revenge" and of the "power of the Earth to kill" as the unforeseen result of how we have been "abusing the Earth" (Lovelock 2007: 1). The dialectic takes the form of a literally deadly payback.

Latour stresses that, contrary to expectations, the natural contract intended in Serres' book title is not a deal between two parties, humanity and nature, but rather is "a series of transactions in which one can see how, all along and in the sciences themselves, the various types of entities mobilized by geohistory have exchanged the various traits that define their agency", *traits* being Serres chosen term to designate the "transactions between the aforementioned subjects and the aforementioned objects" (2017: 64). The idea of an exchange of various traits defining the agency of "various types of entities" is a suggestive one. It seems to suggest that such traits are not in the manner of "once and for all" and to debunk any notion about their being innate properties of distinct entities (individual agents). Indeed, putting it like this reminds us of the view of agency advocated in Barad's agential realism.

Once again, however, it is difficult to be sure about what more precisely Latour has in mind. How does the exchange and the transactions between different entities, impacting on and altering their distinct agency, come about? What sort of entities are we talking about? Saying that they come in human, as well as non-human, forms is not very helpful as far as analytic clarity is concerned. Latour assures us that "Serres is not proposing to *animate* the Earth by claiming that it would benefit from a form of comprehension, sympathy, or sovereignty" (2017: 65). Quite the opposite: "He proposes to take the force of attraction itself as a *bond* that would allow us to understand what is meant by the *force* of law and the *power* of understanding". Accordingly, translation, Serres' overall project, "becomes the way of understanding by what we are *attached* and on what we *depend*. If we become capable of translating, then the laws of nature begin to have a spirit" (ibid.).

There is something paradoxical about Latour's approach. On the one hand, he shares with posthumanism a thoroughgoing critique of anthropocentrism in theory and practice and of the hubris that goes with viewing humans as superior, even unique, in everything to do with agency. On the other hand, he theorizes the agency that is now visibly, and dramatically, operative in the "behavior" exhibited by the Earth in purely human-centered terms, as if humans and their way of understanding what is happening in the Anthropocene are the only thing that matters:

"If we become capable of translating, then the laws of nature begin to have a spirit", as if spirit in the medium of nature were wholly up to humans to create and determine; as if translation were a one-way affair, something we do with respect to something that we are not, that is not us, yet that seems to be lost in translation for lack of exerting its (their) own nonhuman voice(s); as if the animation or deanimation of the Earth is a choice made by humans. Latour's manner of speaking is a far cry from Kohn's formulation that "the world is animate, whether or not we are animists" (2013: 217). But perhaps the expectation that Latour should echo the sort of animism lived by Kohn's Runa people is just mistaken; after all, Latour's major claim to fame is as the champion of a constructivist epistemology whose commonalities with a cosmology such as animism are virtually nonexistent.

Yet we should beware of drawing a conclusion before having looked closely at the steps in Latour's argument. He is certainly on the right track when he observes that we have rendered the material world mute in order to avoid answering the questions "Who or what is speaking? Who or what is acting?" (2017: 67). More specifically, Latour is interested in the operation through which language has seen to it that "some characters are deprived of any form of agency" (ibid.). This sounds correct, although not a word is said about who or what those "characters" are. Forests? Rocks? Brittlestars? Bacteria? Entire ecosystems? This operation in language and by means of language, he goes on, "is going to *deanimate* some of the actors and give the impression that there is a gulf between inanimate material actors and human subjects endowed with soul – or at least with consciousness" (ibid.). Again, fine. But what about the truth claim implicit in giving such an impression? Is it simply untrue that there is the said gulf? If so, why not say so? And if there is no gulf, does that mean that there are no important differences? Are there human subjects with soul – or just languages promoting the impression that there are, erroneously at that (granted that such wrongness may at all be ascertained within the language employed)?

But Latour isn't interested in settling these questions in terms of the validity claims involved. Instead, he makes a series of assertions concerning the role of language, contending that "the idea of an inert world is itself *an effect of style*, a particular *genre*" (2017: 67). "Similarly", he continues, "the idea of a deanimated world is only a way of linking animations *as if* nothing were happening there. But agency is always there, whatever we may do" (2017: 68). Surely the tension crops up again, remaining as unresolved as it is unacknowledged: on the one hand, the formulations that suggest that the matters discussed – deanimation, agency – reside in the eye of the (human) beholder, in keeping with the constitutive role accorded to language and genre; on the other hand, and without a mediation or analytical bridge between them, the blunt assertion that "agency is always there, whatever we may do", suggesting that not only what we think and say through language, but what we do as well, is wholly incapable of determining agency (presumably nonhuman such), so that believing otherwise would appear a form of hubris, thus undermining Latour's own pride of place to language.

Interestingly, the way Latour invokes the role of language seems to suggest that its only practitioners are humans – which, if anything, betrays an unexpected anthropocentric legacy at the heart of Latour's framework for discussing the distinction between animation and deanimation. The subjects of language – the agents making use of it – are of course no less important than its objects, understood as the entities (referents) in the world referred to through language. Admittedly, this is to employ a simplistic, pre-critical, and at any rate pre-constructionist, distinction. Nonetheless, it captures a difference between Latour's cited formulations and, say, Kohn's Peirce-inspired semiotic theory of language, allowing for the appreciation of nonhuman language agents in addition to human ones.

However, the problem I see with Latour's argument goes much deeper, and it has to do with the very situation that Latour is trying to comprehend in *Facing Gaia*. Again, the blunt assertion – without argument – that "agency is always there, whatever we may do", is the key.

Recall the major thesis in Serres' *The Natural Contract*, approvingly quoted by Latour: "For, as of today, the Earth is quaking anew: not because it shifts and moves in its restless, wise orbit, not because it is changing, from its deep plates to its envelope of air, but because it is being transformed by our doing. [. . .] We are disturbing the Earth and making it quake" (Serres 1995: 86). Thus, what we do matters greatly, matters more than ever in history, matters even to the point of profoundly *modifying* – interfering with, damaging, or reinforcing (as the case may be in each particular instance) – *the fundamental capacity for agency among all sorts of actors*: a modification brought about by the actions that humans have carried out, or have helped set in motion, during the last century in particular. The insight into the power of human-initiated activities on Earth to modify the various agencies operative on Earth is what Serres' book and Lovelock's so-called Gaia theory have in common.

I grant that there is another way to interpret his assertion – namely to the effect that "agency is always there" *regardless* of attempts on our part to deny it. If that is his intended meaning, however, Latour should have said, "whatever we may think about it" rather than "whatever we may do". Our doing is precisely the point, the difference that makes a difference to the Earth and all kinds of agents that are part of it, for the joint Serres-Lovelock reason that was just mentioned.

Now Latour's stated interest is in the ways in which "events, adventures, exchanges of properties, transactions among agencies are multiplied from one moment to the next" (2017: 68). However, a deeper understanding of this ongoing multiplication of transactions among various agencies is prevented, and has stubbornly been so for a long time in Western thinking, in that "we distribute agency among causes and consequences in such a way as to attribute everything to the causes and nothing to the consequences, except the property of being traversed by the effect without adding anything to it" (2017: 71). He continues: "We gain access to materiality when we reject this secondary operation that eliminates agents and when we leave the consequences with all the *agency* of which they are capable. It is through the causalist narrative that this effect of deanimation is obtained, but always *after the fact*, once agency has been redistributed among the long series of consequences, once this series has been retooled, set up, and traversed *in reverse order*" (ibid.).

It certainly sounds prudent to stop "eliminating agents" and to "leave the consequences with all the agency of which they are capable". But again, who are they? And where? In the context of the passages quoted, Latour doesn't say: he does not offer examples, cases, or studies that would allow us to better understand what he is referring to and therefore help us to determine whether his account is valid. In a later lecture, however, he takes up the topic again, telling us that "an agent, an actor, an actant, by definition, is *that which acts*, that which has – is endowed with – agency", only immediately adding the question "How can one render the entire world 'inanimate'?", and answering that "this is not a mystification but a *mystique*", though one undermined, it would appear, by the fact that "every discipline, every specialty, every laboratory, every expedition *multiplies* the surprising agents of which the world is made – agents that can easily be followed through the proliferation of technical vocabulary that pervades scientific articles" (2017: 163). The present situation, then, is one where "scientific literature constantly multiplies the number of agents that have to be taken into account in order to follow a course of action to its endpoint" (ibid.). This is indeed true, and it has become increasingly so since Hans Jonas made the same point in his book *The Imperative of Responsibility: In Search for an Ethics for the Technological Age*, first published in Germany in 1979. What the scientific disciplines currently help us appreciate, according to Latour, is that there is "an immense *increase* in what moves, acts, heats up, boils over, and becomes complicated – in sum, in what actually *animates* the agents that constitute the world and in the continuous refinement of the *metamorphic zone*" (ibid.). The upshot of this development, distinctive of the Anthropocene, is that "matters of fact" have become so many "matters of concern" (2017: 164). Scientists of all sorts, especially those working interdisciplinarily, are accorded a crucial role in this.

II

This is where things become complicated. On the one hand, there is Latour's assertion that there is presently an immense increase in what actually animates the agents that constitute the world, this being so as a matter of fact. On the other hand, the fact is that, to the extent that this increase is a "consequence" of human-initiated activities, the brought about situation is precisely *not* perceived as a "matter of concern", if we are to believe what Latour tells us one lecture later. "[T]he notion that we are the agent of the sixth extinction of terrestrial species evokes only a jaded yawn"; hence, "reactivity and sensitivity are what have to be reconsidered". And he adds: "Collectively, we *choose* what we are sensitive too, what we need to react to quickly" (2017: 191).

On the face of it, this may well be true: we are free to be sensitive to the species we, collectively, are now about to make extinct; yet we choose to be insensitive. That Latour should, without argument or qualification, subscribe to the familiar notion of voluntarism liberal style is, however, surprising. That said, the psychology at work here is not that complex or perplexing: perhaps it is precisely *because* we are responsible for the extinction that Latour talks about that we remain insensitive in the face of the increasing pile of well-documented facts about it. If our freedom of choice is such as to allow us to deny uncomfortable facts, or to avoid cognitive dissonance, that may well be what we – or a large majority – choose to do (see Weintrobe 2013 for a psychoanalytic perspective on this).

But these are superficial observations in that there is something superficial, or, more to the point, fictitious about the freedom of choice appealed to. At a deeper level, the option to be sensitive to "the reactions of beings of a certain type", namely nonhuman such in all their variety, is precisely *not* as available to us as its opposite, remaining insensitive, sticking to the insensitivity that, collectively, has facilitated as well as helped justify the practice of extinction, in which we are all somehow implicated in the first place. There is no symmetry between the two alternatives; there is only deep-seated, long-established and structurally sustained asymmetry. And crucially, the "choice" made between them is not a neutral or quasi-agnostic matter but a strongly normative one charged with affective involvement on the part of the individual agent as well as on the part of all the "social others" that will be keen to evaluate, and respond in kind to, the choice made.

To appreciate that this is not first and foremost a theoretical issue, but an eminently concrete one, consider the following. At 4 years of age, my youngest son used to greet the trees, the creek, and the lake where we had our rest during our Sunday walk in the woods. At 6, he stopped this practice which may well be called animist. At 8, being asked by his father why he quit his earlier expansive greeting practice and had limited it to fellow humans only, adopting an observing and objectifying as opposed to a participatory-communicative stance toward the trees, a one-way as opposed to a two-way, reciprocity-oriented one (to employ Habermasian terminology), his answer came promptly and with more than a hint of irritation: "*Of course* I have stopped greeting trees, dad. It is a childish thing to do. I've never seen you, or any other adult, do it. And if one of my classmates should catch me doing it (*in flagrante delicto*), it would be unbearably embarrassing, the talk of the day at school. So why do you ask? Get real, dad. It's such an obvious thing that I no longer do this".

What should attract our attention here is that for *the 4-year-old* it was an obvious thing to do: the greeting of trees and of other "beings of a certain [nonhuman] type" encountered in the woods and spontaneously, without any instruction or role-imitation, related to as persons in the minimal sense of being appropriate addressees of greetings and goodbyes. So what is more conspicuous and in need of explanation: engaging in a practice of greeting trees, showing them respect and being sensitive to them, or not? In my son's response to my "stupid" question, he clearly articulates that for an 8-year old boy in a contemporary Western society, greeting trees, and risking being spotted by others doing so, is no option. That "we choose what we are sensitive to" (Latour) is simply untrue on this concrete, individual, one-to-one (boy-to-tree) level. He may be in the woods, alone with his father. But for all that, he's not acting in a cultural, social or moral vacuum, nor in an epistemic no man's land.

To be exact, Latour's point is that we "collectively" choose. But my point still holds: For the individual being part of that collective, where and how to draw the distinction between entities suited for respect (or sensitivity, as in the display of concern) on the one hand, and those not suited, on the other, is a matter (pre-) decided by the collective whose member one is, and wishes to remain. It is not up for grabs, for the particular individual to decide (see Vetlesen 1994: 193ff.). Animism as practiced, say, in the form of greeting trees and viewing them as persons, alongside and yet in many respects different not only from humans but from frogs and creeks as well, is – for all sorts of historical and cultural reasons – not an option in our society. Whether this fact undercuts the case for the relevance of animist cosmology in our society today, or, to the contrary, underscores its validity and urgency, is a big issue in its own right, one to which I will return below. Suffice it to stress here that our society is one that expects its members to draw distinctions of all sorts (factual, evaluative) in *one particular way* rather than in others; a collectively issued prescription, the successful internalization of which is proven in the individual's everyday behavior, will be rewarded, whereas lack of compliance, not to mention cases of outright resistance, will be duly sanctioned.

All of this is amply taken on board in – as – the mentality of my son once old enough to leave the childish ways of the 4-year-old behind him. Culturally and cosmologically – collectively in a strong sense – we "moderns" hail the shift from greeting trees to not-greeting them as evidence of progress, as a product of Enlightenment philosophy and the scientific explanation of the natural world. As Neil Evernden remarks, as a society we have for the last four hundred years adopted a stance which defines humans' relationship to nature out of existence: "We have denuded the world of subjects and thus foreclosed any possibility of reciprocity or relationship" with the creatures making up the nonhuman natural world (Evernden 1985: 135). The effect is to render any attempt by, say, a boy of 8, not to mention an adult of 50, to *go on* practicing greeting the trees, and in that sense to enact an animist cosmology, into a sign of regression, a conspicuous piece of behavior to be explained as well as handled by means of psychiatry, representing deviance: an idiosyncratic and irrational stance in the midst of a self-proclaimed rationalist, science-driven society.

So, when Latour asks "Why do ecological questions not seem of *direct* concern to our identity, our security, and our property?"; why, that is, the enormous contrast between how we react to "the slightest terrorist attack" and to the sixth extinction of terrestrial species (2017: 191): the questions ring overly rhetorical. The answer is obvious – as Latour admits later on when he talks about how "the deanimation of the world [was] imposed by the epistemological view of scientific activity" (2017: 206f.). Drawing on Eric Voegelin's history of political theology, Latour writes that "for those who have immanentized Heaven, there is no longer any accessible Earth", adding that "the whole paradox of modernization is that it has lost sight, more and more, of any contact-with-the down-to-earth, with materiality: it no longer sees anything in this world below but the other world simply immanentized"; so that, if they "miss out on the world, these Moderns", they do so "from an overdose of ill-placed transcendence" (2017: 200).

Yet the answer to Latour's questions may be more mundane, less lofty and theological: contrary to his suggestion, ecological questions *are* of direct concern to our (or to the large majority's) identity these days, only the way they are so is predominantly, if not to say wholly, negative, being played out in a manner that is inverse and even perverse (I say this without mentioning, or having to mention, Trump). Let me put it like this: our identity is connected to, and to a large extent hinges on, a culturally upheld shared sense of being *right*, because entitled, to treating the myriad of terrestrial nonhuman species the way we, structurally, institutionally, and collectively, do, namely as mere means to human needs, interests, and projects. It is precisely *because* our identity and our self-perception are so heavily involved and invested in this exploitation – now leading to extinction – being something we are entitled to, something that *is worth it*, come what may in terms of destruction of the natural world, something that expresses a "truth" about *who we are* as distinguished from all those "others", namely that we are superior, they are inferior, that we are ends-in-ourselves, they are mere means (Kant), that our sense of identity de facto is inextricably tied to issues of ecology, paralleling as it were the accelerating pace at which the killing off and extinction of nonhuman species is currently happening in the world, anthropogenically so, this being what the Anthropocene is all about. The experience is made, is confirmed, over and over again in our society: few if any issues is as contentious, as explosive in terms of emotional investment and emotional response, to the point of outright aggression, as that of human-caused destruction of nonhuman nature. The more people that have invested in the twin-notion of human exceptionalism and our entitlement to exploit nature, the stronger the need to be unrepentant and defiant in the face of its factual costs, of well-documented looming disaster; the need to deny outright that climate change, biodiversity loss, and species extinction are so many wake-up calls alerting us to the gravity of the situation, the profundity of the ways in which we culturally and collectively have erred. The stronger the evidence, the more important to resist its message; and the longer and more stubbornly one has thus resisted, the more painful the the thought of realizing, and admitting, having been wrong all along (see Dodds 2011; Fisher 2013; Weintrobe 2013). Tellingly, the emotionally charged responses that I have in mind are not directed at *the destruction* (be it for the sake of the species affected, or for the sake of humans affected by their extinction) but at those who point to, or otherwise serve as chroniclers and reminders of, the fact of that human-caused destruction: whistleblowers, critics, and politically outspoken experts such as James Hansen and Michael Mann. It is more important not to rock the boat than to change its steady course, however disastrous.

Return now to the question about what Latour is talking about when he writes that it is "as though we had decided to remain insensitive to the reactions of beings of a certain type – those who are connected, broadly speaking, to the strange figure of matter", prompting the statement that "what we have to understand is why we are not true materialists" (2017: 191). Presumably, then, true materialists would be people who are sensitive to beings connected to "the strange figure of matter". But this is not how a materialist position in philosophy or metaphysics is normally understood. Sensitivity and matter are more commonly seen as antidotes than as exhibiting affinity to each other. That said, Latour could of course reply that this is precisely the problem, calling for a different understanding.

What does that mean? For a start, that what we identify as "matter" *does* invite, even call for, as the appropriate stance to it, a sort of sensitivity on our part, as opposed to the indifference long prevailing in our culture. A true materialism, then, would be an affirmative such, one promoting that we care for, and care about, what is considered matter. But how can you abandon the indifference toward matter which is so much part of our socialization, our culturally induced, and inter-generationally (adult-child, teacher-pupil) transmitted mentality? Well, by abandoning the notion that *matter as such is indifferent*, is flat, mute, is deaf, dumb, and blind, and so inviting on our part indifference because indifferent to the way that we perceive it and what we, based on that perception, do to it. In a word, abandon the long-cherished, long-practiced constellation of *subjects being indifferent to indifferent objects*, because they are indifferent; being insensitive to what is considered itself devoid of anything resembling, and resonating with, our sensitivity. Declaring null and void, untrue the facts, the long-standing deanimation of matter, then, would mean to strike at the root of the constellation: to (re)animate matter, recognizing its own peculiar point of view, or better, series of different, particular such, lest we generalize (recall Whitehead's warning against the fallacy of misplaced concreteness) what is in reality always particular, situated, a someone or something being in the world at some time and some place, constituting in each case a point of view in/on the world, a stance upon things in the world, including us and our actions. Concomitantly, on our part it would necessitate the adoption of wholly different stance, in theory and practice, than the familiar one to undifferentiated "matter" as a perfectly passive and insentient reservoir for human-related ends. Thus understood, a "true" materialism would put anthropocentrism old and new to an end.

This is not Latour's elaboration. It is mine, my go at what it could look like, for lack of his. Again, that lack stems from his vagueness concerning matter, conceptually and as far as matter's empirical manifold and multiplicity are concerned. What is in matter, part of it, what is not?

Grappling with these issues, Latour keeps referring to Voegelin, as if that would help him reach a higher degree of precision and concreteness: Voegelin, for whom "the Moderns have not been secularized . . . but, conversely, immanentized", the "inevitable result" being that "they have no sort of possible contact with the terrestrial, since they can see in it only the transcendent, which would be trying awkwardly to fold itself into the immanent" (2017: 204). But, here as elsewhere, a less lofty and abstract way of getting at what distinguishes matter is called for, so as to clarify, for instance, Latour's assertion that "matter is materiality plus (I mean to say minus) immanentization" (2017: 207). I cannot help but notice how, in Latour's discussion, the relationship between *explanans* and *explanandum* is constantly oscillating, the burden of proof being shifted from one entity to the other, changing from one passage to the next. Latour moves back and forth between talking about terrestrial species commanding our sensitivity and "the activity, the autonomy, the sensitivity to our actions, of the materials that make up the critical zones in which we all reside", raising anew the question why we "seem incapable of responding to the agency of these materials" (2017: 207). I am at a loss to know what the materials Latour here refers to are; he does not say. Since we are left in the dark about what sort of entities go into these materials, it is impossible to assess their powers of agency and so what kind of response to them would be appropriate. Is Latour postulating agency in matter *per se* and *in extenso*, all across the board, as common to all its instantiations? Does everything that exists, whether qualifying as matter or not, exhibit agency in the sense intended, but not clarified, by Latour? Are there at all agency-free areas and ontological regions, or is agency – as potentiality if not as actuality – coextensive with reality as such? Would he concur with Tim Ingold's statement that things are in life rather than life in things?

Instead of the elucidation and differentiation that are analytically required with regard to agency no less than to matter, what Latour offers are variations of some grand narrative or other and the historical shifts between them. His penchant for sweeping generalizations provides little in terms of particular context and particular agents. The deanimation he is concerned with has a "religious – and, more precisely, apocalyptic – origin"; "it results from the narratives of causality that attribute all action to the cause [. . .] and all passivity to the consequences", as we saw above, amounting to an attempt "to empty the world, as much as possible, of any activity" (2017: 208). "Hence the extreme resistance to taking into account the Earth's activity on the part of those who look at materiality as something inert and passive, and who believe that the world they live in is made up only of objects, of simple matters of fact caused by other equally inert matters of fact" (ibid.).

I am probably not alone in noticing how dependent Latour is on invoking, repeatedly, the very dichotomies he proclaims that he wants to leave behind. It is as if he does not think he can make his case against them without them. And as for leaving them behind – what would that mean? It might make for a night in which all cows are grey; the suggestive yet unspecified proposal for a scenario where everything that used to be heaped together as "matter", as indistinct instances of inert and inanimate such, now, at long last, is allowed to emerge as so many agents, so many instantiations of agency, being granted by us the very properties we, not since time immemorial but during the last four-hundred years or so, have self-servingly deemed ours alone, setting us apart from everything and everybody else on earth. There is a pathos of liberation to such a scenario, a kind of "set all the prisoners free, let them flourish, it is spring".

III

It is instructive at this point to take a look at Jane Bennett's much-discussed book *Vibrant Matter: A Political Ecology of Things*, hailed as a "manifesto for a new materialism" (Rita Felski, blurb). Bennett's approach is strongly influenced by Latour, whose attempt to "describe a more *distributive* agency" she admires (2010: viii).

The stated aim of Bennett's book is to shift focus from the human experience of things to "things themselves", allowing us to recognize and theorize what has for so long been neglected in philosophy and political theory (Bennett's field), namely the active participation of nonhuman forces in various events. What we need to grasp, then, is the "vital materiality" that runs through and across bodies, both human and nonhuman.

To challenge the dominant image of dead, intrinsically inanimate, and thoroughly instrumentalized matter, Bennett takes up the notion of a "material vitalism" where "vitality is immanent in matter-energy", advanced by the French philosophers Gilles Deleuze and Felix Guattari. Bennett's vitalism, she tells us, is not the traditional kind. Drawing on Spinoza, she equates affect with materiality, "rather than posit a separate force that can enter and animate a physical body" (2010: xiii). In the resistance to anthropocentrism that her vital materialism is meant to represent, pride of place is given to "the agentic contributions of nonhuman forces (operating in nature, in the human body, and in human artifacts)" (2010: xvi). Bennett presents her project like this: "If matter itself is lively, then not only is the difference between subjects and objects minimized, but the status of the shared materiality of all things is elevated. All bodies become more than mere objects, as the thing-powers of resistance and protean agency are brought into sharper relief" (2010: 13). Rejecting the dualism between humans and nature, Bennett maintains that "humanity and nonhumanity have always performed an intricate dance with each other. There was never a time when human agency was anything other than an interfolding network of humanity and nonhumanity; today this mingling has become harder to ignore" (2010: 31). The vital materialism Bennett champions takes the form of "a theory of relativity (of sorts): the stones, tables, technologies, words, and edibles that confront us as fixed are mobile, internally heterogeneous materials whose rate of speed and pace of change are *slow* compared to the duration and velocity of the human bodies participating in and perceiving them. 'Objects' appear as such because their becoming proceeds at a speed or a level below the threshold of human discernment" (2010: 57f.).

For all her declared debt to Latour, the Frankfurt School – Adorno in particular – is a major influence in Bennett's description of how we have come to turn matter into so many instances of passive objects for us to manipulate and control, exploiting what are regarded as mere means with impunity. Unfortunately, in Bennett's case for a vital materialism, one that aspires to do justice to the agency operative in matter, whatever progress is made over traditional, Cartesian-inspired notions of matter and its properties comes at a high price. What Bennett fails to differentiate: everything everywhere is an agent or actant (Latour), is partaking in everything that comes to pass, and so is described in one and the same manner, all across the board. It is Adorno's "philosophy of identity", identitarian thinking, the one he warned against, only this time around the casualties are not conceptual ones (entities) but are eminently material, physical ones.

How can you describe, in like manner, phenomena as different as "stones, tables, technologies, words, and edibles"? What is gained by doing so? What is the argument to show that what these entities (allegedly) have in common, in terms of agency, is so important as to justify silencing their differences, hence their respective particularity? What does "nonhumanity" refer to in the statements about the dance between, and the interfolding networks of, humanity and nonhumanity? Is it simply everything that is not part of humanity? One is hard-pressed to come up with candidates for more broad and crude categories, and the decision on Bennett's part to avoid making distinctions may carry the torch of posthumanist critics who see symbolic violence, illegitimate hierarchies and subtle suppression whenever and wherever "traditional philosophy" draws them. But the gain is as illusory as the categories are vacuous.

Toward the end of her book, Bennett suggests that her vital materialism allows her to be more radical than positions that, invoking the idea of "second nature", hold that what we used to call natural is actually the cultural determination of nature. Bennett's position advances the more provocative view that "culture is *not* of our making, infused as it is by biological, geological, and climatic forces"; after all, she reminds us, there is "a life of metal as well as a life of men" (2010: 115). What we need to accept, then, is that "nonhumanity infects culture". We need to admit the "blasphemous idea that nonhumans – trash, bacteria, stem cells, food, metal, technologies, weather – are actants more than objects", an actant being, as Bennett explains by reference to Latour, "a source of action that can be either human or nonhuman; that which has efficacy, can *do* things, produce effects, alter the course of events" (2010: 115, viii).

What is explained by this? What is Bennett referring to – in terms of events, effects, forces? Again, if everything is a part of everything everywhere, the point is indeed made that nothing can be, should be, excluded, on pain of overlooking something important. However, that is a largely conceptual and as such programmatic point, nothing more; nowadays a rather unoriginal and moot one, at that. So what about the positive point, that of showing what exactly contributes "agentially" to what exactly?

Bennett credits Guattari with acknowledging, as she says Latour does too, "the porosity of the borders between subjectivity, society, and machines", the point being to "reject any attempt to unstir the cream from the coffee – to disentangle the cultural from the natural". She goes on to quote, approvingly, Guattari's statement in 1986 that it makes no political sense to "try to withdraw from nature, for the health of the planet is increasingly reliant upon human intervention, and a time will come when vast programmes will need to be set up in Earth's atmosphere" (2010: 116). One is left wondering what Bennett, following Guattari, is advocating here, relevant to the Anthropocene. Geoengineering? That would certainly fit the bill as far as "vast programmes" are concerned. On the other hand, it would be more than ironic if a book that so explicitly sets out to target the hubris of anthropocentrism and instrumental reason Western style ends up with no other idea about how to tackle our present predicament – ripe with Timothy Morton's quasi-autonomous "hyper-objects" – than the most hubristic, capital-driven and grandiose of them all, as documented in Clive Hamilton's *Earth Masters: The Dawn of the Age of Climate Engineering* (2013) and Naomi Klein's *This Changes Everything: Capitalism vs. the Climate* (2014: 256–290).

On a more philosophical note, Bennett would do well to follow Tim Ingold's advice, and focus attention on the vitality of *materials* – different such, each with their distinct properties, being situated in different contexts including those of time and space, moving and/or being moved in the larger world – instead of speaking about the agency of "matter", of "nonhumanity" and of "nonhuman forces", failing to specify what either of them is comprised of.

In a passage criticizing Latour's notion of "hybrids", the points that Ingold makes are applicable to Bennett as well:

> Air and water are not entities that act. They are material media in which living things are immersed, and are experienced by way of their currents, forces, and pressure gradients. That is, it is not the butterfly alone that flies but the butterfly-in-air, and not the fish alone that swims but fish-in-water. But that no more makes the butterfly a fly-air-hybrid than it makes the fish a fish-water-hybrid. It is simply to recognize that for things to interact they must be immersed in a kind of force field set up by the currents of the media

> that surround them. Cut out from these currents – that is, reduced to objects – they would be *dead*.
>
> (Ingold 2011b: 92f.)

"What academic perversion", asks Ingold, "leads us to speak not of *materials and their properties* but of the materiality [and agency] of objects?" What we need to understand, and what would represent an important task for a "new materialism" worthy of the name, is what Ingold calls "the material composition of the inhabited world", a world of all sorts of different materials – some man-made, others not; some solid and constant, others easily perishable; some moveable in their own right, others only moveable by external forces, still others not at all – that we can only gain access to *practically*, by exhibiting our own peculiar capacities of movement, of engaging with and of taking up a position and being a particular point of view in/on the world that we co-inhabit with so many others whose otherness from us we need both to conceptualize and to handle practically. We need, Ingold says, to engage directly with the stuff, the material, that we want to understand. And yet, academically, even in a work hailed as the manifesto for "new materialism" such as Bennett's, materials in their eminent physicality, in their tangibility, have gone missing, being forgotten and hidden from view, as it were, by a discourse on the relation between materiality (unspecified, and making a virtue of the failure to specify) and "a host of other, seemingly unfathomable qualities including agency, intentionality, functionality, spatiality, semiosis, spirituality, and embodiment" (Ingold 2011b: 20, 213).

Since Ingold's position is important in its own right and not just as a critique of Latour and Bennett, let us examine what may well be his most systematic statement concerning agency:

> Our concept of agency must make allowance for the real complexity of living organisms, as opposed to inert matter. It is simply absurd to place a grain of sand and an aphid on the scales of a balance and to claim that they are equivalent. They may weigh the same amount, but in terms of complexity they are poles apart. They key difference is that the aphid, animal that it is, has a nervous system – just as you and I do. When I crouch at the centre of my web, I am all a-quiver, just like the leaf of a tree in the summer breeze. I am sensitive to the slightest movement of vibration. What makes the difference between me and the leaf, however, is that every movement I make

> is also a movement of my *attention*. It is the attentiveness of this movement that qualifies it as an instance of *action* and, by the same token, qualifies me as an *agent*. To put it another way, the essence of action lies not in afore-thought (as our human philosopher would claim) but in the close coupling of bodily movement and perception. But that is also to say that all action is, to varying degrees, *skilled*. The skilled practitioner is one who can continually attune his or her movements to perturbations in the perceived environment without ever interrupting the flow of action. But such skill does not come ready-made. Rather, it *develops*, as part and parcel of the organism's own growth and development in an environment. Since agency calls for skill, and since skill arises through development, it follows that the process of development is a *sine qua non* for the exercise of agency. To attribute agency to objects that do not grow or develop, that consequently embody no skill, and whose movement is not therefore coupled to their perception, is ludicrous.
>
> (Ingold 2011b: 94)

Several issues call for comment in this rich passage. To begin with, it may seem disappointing that Ingold takes as his point of departure the simple distinction between living organisms and "inert matter". Yet his employment, for once, of the Cartesian term should not mislead us: Ingold does not (suddenly) regard "matter" as "inert". What Ingold wants to pronounce on is not matter as such, matter in general and conceived of abstractly; rather, it is materiality, or more precisely, materials in their specificity, having a specific trajectory in time and space in terms of media and impact on/from other materials and other agents. Materials, such as the wood from which a chair or table can be made, are precisely not inert but a medium of change and movement, helping facilitate that of other sorts of agents and their projects in the world.

More important, however, is the move Ingold makes when he elucidates the "difference between me [himself as a human individual] and the leaf", the critical difference being that "every movement I make is also a movement of my *attention*". To be sure, on the face of it a human individual exercises capacities for attention not found in a leaf. But Ingold's way of making his claim is incoherent. A valid comparison would be that between an oak (a tree) and a human individual, or that between an oak leaf and a human's arm. You cannot make the comparison by considering the one entity in its quality as a(n) (organic) whole, and the other as a part, separate from its whole; to do so is to place the relata on different levels, not only logically so, but *ontologically*. Ingold here does to the leaf what he never misses a chance to warn against in others: cut out from the force field and the currents – that is, reduced to an object – the leaf would be *dead*. It is only a leaf, and only able to remain so and do what, say, an oak leaf is meant to do, to be, and to *thrive as being*, as long as it is part of the tree, or better still, a member of the organism that is the tree – the tree that for its part would be no such thing without the (its) leaves.

Perhaps no philosopher has argued this point better than Hegel:

> Through this unity of doubled activity all members of the organism are constantly preserved and constantly taken back into the ideality of their animation [*Belebung*]. The members manifest this ideality [of their differences] in that their animating unity is not something indifferent to them, but on the contrary is their very substance in which and through which alone they can preserve their particular individuality. This fact constitutes the essential difference between a *part* of a whole and a *member* [*Glied*] of an organism. The special parts of a house, for example – the stones, the windows – remain the same whether they together form a house or not. Their community with the others is indifferent to them, and the *Begriff* remains in them a mere external form, that does not live in the real parts in order to raise these to the ideality of a subjective unity. In contrast, the members of an organism likewise have an external reality, but the concept is so much the indwelling essence [*Wesen*] of the organism, that it is not imprinted on them as a form uniting them only externally; on the contrary, it is their only possible mode of enduring and surviving. For this reason, the members of the organism do not have a reality possessed by stones of a building, or the planets, moon and comets in the planetary system; rather, in spite of all their reality, they have an existence posited as ideal within the organism. For example, the hand, when severed from the body loses its independent existence. It does not remain the same as it was in the organism; rather its mobility, movement, structure, color, etc., are changed. It begins to decay and its entire existence dissolves. It has a sustained, enduring existence only as a member of the organism, and has reality only as continuously brought back into the ideal unity. Herein consists the higher mode of reality within the living organism.
>
> (Hegel 1975: 121)

As will not be lost on the reader, Hegel anticipates Whitehead's fallacy of misplaced concreteness, showing Whitehead's debt to Hegel's crucial distinction between *mechanism*, understood as the logic and ontology of externality and exchangeability, and *organism*, understood as that of internal (interdependent) relations and essence (see Williams 1997: 313, footnote 64). Hegel's out-of-date idealist vocabulary aside, the point he makes is spot-on: in separating and severing the leaf from the tree (its tree), Ingold does the leaf the same injustice he would have done to the hand if he separated it from the human mind-and-body organism.

When Ingold states that "the skilled practitioner is one who can continually attune his or her perturbations in the perceived environment without ever interrupting the flow of action", he seems to have only human individuals in mind. But we do not need to read Kohn to realize that his description holds for jaguars and peccaries as well. Whether the tree from which each individual leaf is inseparable is such as to allow the description that "its [the tree's] every movement is also a movement of its attention", is harder to tell on the basis of Ingold's account – the inclusion of the tree seems more contestable than that of the jaguar. Still there is no doubt that the tree qualifies insofar as it grows and develops; in that sense, it is no less an agent than the human, the jaguar, and the peccary. Moreover, as the German forester Peter Wohlleben shows in *The Secret Life of Trees*, drawing on recent botanical research, each tree is a particular point of view on/in the world, stretching out in the direction of the sun and avoiding shadow, communicating with and constantly adjusting and attuning its movements, its growth, and its flourishing "to perturbations in the perceived environment", being one of many members of that environment at that particular time and place and making a difference to it (within it), to paraphrase Ingold.

The bottom line with Ingold is one that he shares with Hegel and Whitehead: never start with a separate, individual entity identified as a mere "object". And likewise, never start with an abstract category such as "matter" (be it to argue the case for it being, immanently, "vibrant", allowing for an ecology of "things", to recall Bennett's book). We never start out in such a manner in practice, as Heidegger (1962 [1927]; Dreyfus 1991) showed in his famous argument for the primacy of *Zuhandenheit* ("ready-to-hand") over *Vorhandenheit* ("present-at-hand") in *Being and Time*; accordingly, we should not do so in philosophy either. Instead, we engage *in situ* and *in medias res* with various, and very different and diversified, co-inhabitors on planet Earth, each of them as concretely *in situ* as we are, rendering the access to them and the effort to gain knowledge about them never something abstract and general (these being the trademarks of what since Plato has been considered the highest form of knowledge, of true science, namely *episteme*), but always something eminently context-bound and particular – recall (from Kohn) the very *precise* knowledge that a hunter needs to have about particular animals at a particular place and time of day/night, such that mistaking one animal for another could mean either being killed or missing out on the game, there being no time to second-guess or change one's mind, so that correct identification here-and-now is a matter of life and death, on both sides of the human-animal relationship.

IV

In *Facing Gaia*, the most important among the dichotomies that Latour sets out to challenge and destabilize, and yet that he can never make his point without, is clearly that between animation and deanimation. Recall his strong claim: "Although the official philosophy of science takes the [. . .] movement of deanimation as the only important and rational one, the opposite is true: animation is the essential phenomenon; and deanimation is the superficial, auxiliary, polemical, and often defensive phenomenon". This being so, the truly naive belief is not that in animism but that in a deanimated "material world" (2017: 70). In other words, Descartes and his followers up to the present in the era of modern science are naive, the Ecuadorian Runa and Brazilian Yanomami are not. Needless to say, it is a deliberately provocative turning of the tables.

What Latour aims to do is to "bring out the gap that separates the common-sense assumption that one can easily distinguish between the objects of the natural world, on the one hand, and the subjects of the human world, on the other, from the extreme difficulty of making this distinction in practice". This leads to the contention that we looked at above, that "the actors, with their multiple forms and capacities, never stop exchanging their properties" (2017: 57). The upshot is that anthropomorphic representations are – or more to the point, become – unstable; they mix with, are modified or displaced by, hydromorphic ones, biomorphic ones, phusimorphic ones (ibid.). So, "if it is the world that interests us – and no longer 'nature' – then we must learn to inhabit what could be called a *metamorphic zone*", a term Latour borrows from geology to capture "all the 'morphisms' that we are going to have to register in order to follow these transactions" (2017: 58).

We saw that Latour has great faith in the role that science – and different types of scientists – can play in not only studying, but also in helping the wider public acknowledging, the myriad of agents and agencies that will play themselves out once the "superficial, auxiliary, polemical, and often defensive phenomenon" of deanimation loses its incapacitating hold. Remove the straitjacket, and all the agencies kept mute burst forward and claim the voice denied them. Their muteness – inertness – was never their intrinsic property, their in-itself, but was always a condition imposed on them by us and for our self-obsessed, narrow, and shortsighted purposes, untruthfully and artificially so, and as we are learning these days: unsustainably so.

The fact, then, is not that "scientists' objects of study are *inanimate*" but that "our degree of familiarity with these objects or 'actors' is very slight; the inanimate 'actors', or *actants*, thus need to be presented at greater length than the characters we call anthropomorphic, with whom we believe we are better acquainted" (2017: 49f.). So what Latour is asking his readers to do is "to suspend the usual reading grid that makes us tend to contrast human and nonhuman actors, for example, subjects and objects". He would like us, he continues, "to remain attentive to what constitutes their common repertory. It will then become clear that to say of an actor that he/she/it is inert – in the sense of having no agency – or, conversely, that he/she/it is animated – in the sense of 'endowed with a soul' – is a secondary and derivative operation" (2017: 50).

Secondary and derivative from whose point of view? Note that everything that Latour is saying about agents and actants, subjects and objects, the animate and the deanimate, is said from one particular point of view: the human one. Yet Latour's declared message is about the many, not the one, not any particular one, especially not, of course, the castigated anthropocentric point of view. Nonetheless, in his way of writing, it is always a matter of what and whom *we* look upon as subjects, or as objects, and with what and whom *we* are more or less well acquainted. And indeed, we do "tend" to contrast human and nonhuman actors, viewing and treating some as subjects, others as objects. *So do nonhumans*, and for good reason, just as do we. A reason for their doing so, a supremely good one at that, is survival: the jaguar, Kohn explained in Chapter 4, needs to be just as apt at humans (in their role as hunters) in distinguishing human from nonhuman animals, and subjects from objects, so as to capture the difference between those who may attack or flee, and those unable to; those suitable for eating and those not. To make the difference, and to make it correctly, in the spur of the moment – here, now, facing *these* others – is not a matter of choice of (electable, optional) "reading grid", or of performing "a secondary or derivative operation"; it is a matter of life and death, the condition for the very being – sustained such – of everything and everyone that knows what it means to act and to be acted upon. Indeed, above I said much the same, only from the human hunter's point of view – the point being that the description given holds for both, humans and animals alike.

To be sure, it is a good thing to become attentive to what constitutes the "common repertory" between beings kept apart by old-worn dichotomies and hierarchies. But Latour never makes the argument from the point of view of any – not a single one – of the nonhuman agents for the sake of whom he, presumably, as a critic of anthropocentrism, sets out to make his case in the first place. The choice of reading grid is one that *we* make, one way or the other, and very loftily and intellectually at that, not nonhumans. Kohn writes from Ecuadorian Amazonas, Latour writes from Paris: what you see depends on where you sit. Though Latour's point is clearly to allow what used to be considered objects to step forward in all their agency as subjects (or subjects of sorts), his prose is such as to never actually allow them to do so, thereby betraying his methodological and epistemological anthropocentrism in the very carrying out of his declaredly anthropocentrism-critical argument.

In his earlier work *Politics of Nature*, Latour proposed to speak of "*actants*, acting agents, interveners", instead of actors, seeking to "rid the word of any trace of anthropomorphism" and to make "a longer list of candidates for action", of "elementary actions", allowing us to "redistribute the cards between humans and nonhumans". Actant is a term from semiotics "covering both humans and nonhumans" (Latour 2004: 75, 76, 237). What Latour aims for is "the extension of the collective", making possible "a presentation of humans and nonhumans that is completely different from the one required by the cold war between objects and subjects". The latter, he contends, "were playing a zero-sum game: everything lost by one side was won by the other, and vice versa". But whereas "objects and subjects can never associate with one another, humans and nonhumans can" (2004: 76). Moreover, "there is no reason to reserve the terms 'speech' and 'social actor' for humans, since they are perfectly suited to the nonhumans with which humans share their collective existence increasingly every day, thanks in large part to the work of laboratories" (2004: 77f.). Holding that what human and nonhuman actors have in common is both more fundamental and more relevant than what may separate and distinguish them, Latour maintains that both "appear first as troublemakers", the notion of "recalcitrance" being the most appropriate approach to defining their action, adding that "anyone who believes that nonhumans are defined by strict obedience to the laws of causality must never have followed the slow development of a laboratory experiment" (2004: 81).

That may well be true. But it recalls one of the critical points that I made about Barad's agential realism in Chapter 3: that the laboratory experiment, despite the rigor it may offer in facilitating the making of "the cut" (Barad), is a poor, ill-suited, and all-too-restricted model when it comes to appreciating entities in all their distinctness and variety as far as capacities for agency are concerned. For most of the time, entities/agencies, both human and nonhuman, are what they are, and perform as such, in non-laboratory kinds of roles and contexts, so that generalizing and extrapolating their agencies from what shows up in the laboratory – be it literally or metaphorically – is a notoriously difficult thing to do (lest you consider the laboratory to be coextensive with reality or the world as such, whereby the clarity sought for slips away once again).

Latour's claim is that "In sharing the competencies of speech, association, and reality among humans and nonhumans, we have put an end to the anthropomorphism of the object-subject division that mobilized all entities in a fight for control of the common world". By seeking to "redistribute" speech, association, and recalcitrance among what used to be kept apart as subjects and objects, humans and nonhumans alike will finally be allowed to partake in a greatly expanded collective, in keeping with Latour's plea to "connect the question of the common world to the question of the common good" (2004: 82, 98).

But again, who is Latour talking about? While it certainly seems laudable to want to go beyond the conceptual discrimination as well as practical exploitation that the old anthropocentric order of words, things, and practices implied, to the detriment of those who only qualified as objects, not subjects, as nonhumans, not humans, what precisely is gained by the expansion being promoted? If the new collective is an all-inclusive one, what and who more exactly does it encompass and include? And what happens to the distinctness, not to mention uniqueness, of each particular member?

V

In one of his rare examples Latour talks about the Mississippi as an agent not an object: a river may do things, cause actions, be at the initiating end (not forgetting the all-important difference between being situated upstream or downstream), not at the receiving one. Surely, a particular river in a particular region has a lot of experience with the difference between what it means *to act* and *to be acted upon*, being capable of both. Mississippi, then, is a "truly *natural* actor", a "force of nature" that is "obviously just the opposite of an inert actor", so that "if the Mississippi possesses anything at all, it is *agency* – such powerful agency that it imposes itself on the agency of all the bureaucrats" (2017: 52). Typically, yet revealingly, it is "the tragedy of Katrina" that has "reminded us" that "the entire Mississippi basin, completely artificialized, is attempting to protect itself behind the fragile front of its dikes" (2017: 53). May we assume that were it not for the ways in which so many *human agents* in New Orleans were affected by Katrina in 2005, the reports Latour is citing concerning the agency and power wielded by "natural actors" such as the Mississippi river would never have received such extensive coverage? More principally, is the existence of human actors, and the point of view taken by them, not to mention the way they are concretely affected in cases of "natural disasters", the operative *sine qua non* for the acknowledgment of agency in "natural" actors in which Latour is engaging?

Let me be explain. In a footnote, Latour tells us that what led him to explore ecological questions was "the seeming paradox in the fact that the so-called question of the environment appeared when the external environment disappeared" (2017: 36, fn. 69). That a crisis with destructive consequences may serve to open access to phenomena that is otherwise, under normal circumstances, beyond our reach and outside our sphere of interest, is a notion well known to philosophers at least since phenomenologist Edmund Husserl's work on the lifeworld. Especially relevant to the case of Latour is the late Australian philosopher Teresa Brennan's take on the so-called crisis motif:

> As the biosphere and species are destroyed, and human quality of life is degraded, the owl of Minerva points to the natural laws governing and regulating the survival of what has been or is being destroyed. *One learns what a necessary condition of life was as that condition ceases to exist.* This diabolic experiment with the earth and its creatures reveals how individual human organisms depend like other organic life on the diversity of the whole biosphere.
>
> (Brennan 2004: 161; my italics)

Brennan's observation, intended as a warning, a wake-up call for action, appears at the end of what was to be her last book, *The Transmission of Affect*; it was penned just before she died in a car accident. That the environment, following Latour, became a topic in its own right only at the moment of its destruction is no more "paradoxical" than that Habermas, inspired by Husserl, felt able to lay out the symbolic structures of the lifeworld only at the moment of their being "colonized" by the imperatives of the market and state systems, causing a series of pathologies within the domains of society, culture, and personality – respectively, anomie, loss of meaning, and various psycho-pathologies. As Habermas puts it: "It may be that this provocative threat, this challenge that places the symbolic structures of the lifeworld as a whole in question, can account for why they have become accessible to us" (1987: 403). The idea is that only when something breaks down do you start grasping its till then taken for granted importance, unnoticed because intact and hence enabling you to concentrate on other things. Only qua endangered does a condition on whose intactness we have tacitly counted command our attention – often too late at that.

Enter the Anthropocene to illustrate Brennan's point. The crisis addressed in Latour's *Facing Gaia* could not have been of a greater magnitude. It includes: the extinction of species unprecedented in pace and scope, the loss of biodiversity, the shrinking of habitat available for nonhuman creatures and life-forms of all kinds, the acidification and plastification of the oceans, the melting of glaciers and the release of methane from permafrost and from the seabed, the dying of coral reefs, the rising temperatures and sea levels – to name but a few instances of the crisis set in motion. By humans. Humans who have, for the most part, reinforced rather than halted the crisis since being alerted to it. Of the nine critical boundaries/thresholds of the Earth system that have been identified by scientists, three have already been crossed (climate change, biodiversity, and human interference with the nitrogen cycle), representing extreme rifts in the Earth system, while others – ocean acidification, global freshwater use, changes in land use, and the phosphorus cycle – represent emerging rifts (see Magdoff & Bellamy Foster 2011: 12ff.; Angus 2016). Specifically, climate change does not occur in a gradual way, one obeying simple linear causality, let alone monocausality, but rather takes the form of tipping points fed by amplifying feedbacks that can hasten change and greatly exacerbate and proliferate its consequences, consequences whose onset and reach, impact and devastation, become not only difficult but all-nigh impossible to foresee in a precise manner.

What is the lesson? We not only took nature for granted in the phenomenologists' sense, revealing a deep-seated, quasi-transcendental and as it were ahistorical propensity of ours, presumably to be found in all human agents at all times. Rather, eminently shaped by specific historical and cultural factors, we came to take nature's intactness for granted, and its continued such. In a complacent manner bordering on sheer carelessness and recklessness, it now slowly, reluctantly dawns on us; taking nature for granted even as we go on over-exploiting it, wrecking it. The super-efficient technologies that presently facilitate, reinforce, and accelerate that over-exploitation by way of systemic overshoot (overstepping limits, taking without allowing nature to reproduce and replenish), have stunted our ability to notice the destruction we are busy causing. The natural world has become abstract, remote, bracketed, a quasi-separate, eerily "autonomous" sphere (Merchant 2016) beyond our daily arenas, experiences, and projects, these being increasingly culture-related not nature-related ones, closed in on themselves and the universe of "social" (fellow human) others, not open toward and interacting with "natural" (nonhuman) such, the latter being out of sight, out of mind to an extent totally unknown to nonmodern cultures and life-forms (Vetlesen 2015: 147–192; Skrbina 2015). Collectively, as a culture, we have acted as if nature, considered and treated as matter in the Cartesian, narrowly materialist sense alluded to by Latour, were, in whole and parts, replaceable, disposable and dispensable.

But it is not. The error that we made and that we go on committing is a *factual* one: one of misidentification that in its turn justified mistreatment. That's what happens when something – be it as a part (Hegel: member), be it as a whole – that possesses value in itself and helps (re)produce and sustain value in the wider system that it is part of, is treated *as if* it a mere means: it has consequences. We have acted toward nature – understood as the totality of what is other-than-human – as if it were up to us (part of our freedom) to decide whether or not there is intrinsic value to be found there. We as a culture decided there is not, and acted upon that stance, viewing our capacity to do so as proof of our superiority over and above the entire world of nonhuman creatures. But in reality we are not in a position to decide about the presence of value in all things existing; it is a role, based on a sense of entitlement, that we have simply granted ourselves, consulting no others. What we do possess, however, is the very opposite power from that of construction, of bringing about and conferring, licensing, (all) value, namely the power to destroy: to spoil all those nonhuman entities in which value resides, or used to reside, until we treated them as means only and thereupon thwarted their intactness qua ends in themselves, their species-specific ability to flourish and replenish according to what they are and what they are meant to be in each particular case. Revealingly, and if at all, the only way in which such a culture can conceive of the destruction inflicted, is in terms of "loss of ecosystem services" – as if the whole of nature is a commodity market and the only roles that matter are those of buyers and sellers, producers, and consumers.

So the paradox may be a seeming one, as Latour calls it. Instead of a paradox, or something that should take us by surprise, it is the old story all over again, only this time around much graver than before: not a local or temporary matter, but an all-encompassing and irreversible one. We humans are not very good at spotting, and reacting to, human-caused dangers in the natural world, be it that the danger that the natural world as we know it – used to know it – collapses. The philosopher Martin Lee Mueller puts it well: "What becomes of this multiplicity, this living diversity, when a single mode of upsurge seeks to monopolize all others?" Indeed, "how can a truly modern people really think that rational thought cut off from its source will not eventually wither alongside all those others who are now withering, or who have already slipped away?" (Lee Mueller 2017: 99).

The paradox – and there is one – in the case of Latour is this: He gives the impression of being strongly aware that he is writing about Gaia at the moment it's in peril. Its being in peril may even be considered his reason for writing about it, for finding it interesting, a matter of concern, not just a matter of fact, to cite the formulation that he has been so fond of ever since he wrote *Politics of Nature* (2004). And yet, the manner in which he writes about it is oddly out of sync with the reality of the situation, the gravity of which being the message he purports to take over from Lovelock. Actors, actants, all sorts and variations of agencies long denied, subdued, ridiculed by "one creature among others", focused only on the "singular importance of its own, rational attunement to the world" – to cite Lee Mueller (2017: 99) – all these "deanimated" beings, creatures, entities that go to make up what we call nature, what we treat as matter, are somehow – as indicated in Latour's often poetic prose – energized, bursting forth, throwing off the straitjacket that our "false" materialism has imprisoned them in, playing themselves out presently, everywhere, before our largely unprepared and uncomprehending eyes. As a reader, one cannot help but be fascinated by Latour's fascination with this explosion of agents, actants, and agencies that we, contemporaries to this flood of nonhuman forces, inundating our societies and lives, are experiencing.

Why "out of sync"? For the simple reason that we have to do with *loss*, at unprecedented pace, of all sorts of nonhuman agents, actants, and agencies, not their proliferation. Diversity – cultural as well as biological such – is being drastically reduced, more so each year and as we speak. Monoculture is displacing manifold and richness in all dimensions and on all fronts. Whereas the number of us – humans – steadily increases, the number of all sorts of others decreases: not coincidentally so, but strictly causally, with the one leading to the other. Humanity's metaphysical homelessness in the world – to allude to post-WWII existentialism à la early Sartre and Camus, insisting on the "absurdity" and "futility" of man's being-in-the-world – is overshadowed by an altogether different one: our eminently non-metaphysical, since self-inflicted, biological, species-related loneliness on planet Earth. Virtually everything a present-day child sees, hears, smells, and touches is human-made, artifacts for human purposes, commodities produced, distributed, bought and sold by means of human-engineered technologies. If anything, the diversity and multiplicity of agents and agencies in the world – apart from the human-centered ones just mentioned – is shrinking, their habitats fragmenting, disappearing; richness is giving way to poverty, complexity to homogeneity. In the last forty years, the number of animals in the world is halved. And so on and so forth.

Latour surely knows this as well as I do. Yet the truth of it and its stark impacts are absent from his *Facing Gaia*. It is more a matter of evading Gaia than facing her.

Consider, again, Latour's fascination with the agential potency of the Mississippi river in the event of the Katrina hurricane in 2005. Or any other "extreme weather" event during the last decade or so, showing up so many "natural disasters" that, indeed, are often seen as reminders of how potent an agent nature is – still is, despite our efforts as Civilization, as a science-driven, technological society to "tame" nature, to fully manipulate, modify, and control it so as to serve our needs and suit our goals, no such being recognized in nature in itself. As I just remarked, we now – belatedly – realize (or defiantly refuse to realize, for that matter, not that it matters much) that this amounts to a grand *factual error*, following from a set of very specific Western epistemological and ontological premises (reviewed in the above chapters); the most consequential error in the history of mankind, indeed in the history of planet Earth, to use words big enough to rise to the occasion.

The irony that I see in Latour's fascination with a river like the Mississippi in a situation like ours, when Gaia (following Lovelock) is destabilized and its equilibrium in peril, is that his newfound fascination to the point of quasi-"discovery" of the agency of a river concerns something that is not novel at all. Quite to the contrary: nothing is older in the history of Earth, and in that of mankind as well, than the acknowledged fact that, say, a river is a potent agent. Animists of all stripes, through the ages and across the continents, have always known, responded to, and acted upon *the reality of that agency*, being a *sine qua non* for human existence, an existence not to be had and not known without it – only we no longer know, or want to know, and this long-reigning not-knowing is what allows Latour to present his analytical ascription of agency to a nonhuman agent like the Mississippi as a philosophical novelty, as a bold and fascinating thesis now put forward by the father of actor-network theory and the like. To the Kuna and the Yanomami, nothing could be more misleading, verging on a blatant hoax, than intimating a novel discovery here: they would immediately call the bluff in selling this insight as the new intellectual commodity on the global market. The agency of nonhuman entities such as a river, for them, and for virtually everybody else up until four hundred years ago or so in the Western world, is cosmologically as well as experientially elemental, an indubitable reality if ever there was one, all-important for all practical purposes, ultimately for sheer survival. *Every child knows* – but only for as long as we, representatives of the enlightened, science-based adult world-view, allow them to know. As the example of my son shows, it is a knowledge that is quickly and effectively unlearned by cultural means – we call it successful socialization.

But, it may be objected, does this not miss what is truly *novel* in the situation addressed by Lovelock and Latour – namely, that the Earth *responds* to our doings insofar as they are destructive to its functioning? But again, this is precisely not a novel idea in the history of human societies. Recall, from the ethnographic studies that we looked at above, the role of the shaman: to negotiate the relationship between human actions on the one hand and the other creatures and forces making up nature on the other; to see to it that no members of the tribe act so as to upset what is perceived as a precarious balance between the different agents and forces in question, acknowledging that humans are the inferior party in a relationship of dependence; and to show regret and ask for forgiveness in cases where the proper boundaries have been overstepped by human action, spurred by human failings like greed or vanity. To be sure, the degree to which this balance can be upset today, in the era of systemic overshoot and over-exploitation, is incomparably higher than in any previous era, where the ecological problems caused by, say, excessive hunting would be local and limited, and therefore makes for an incomparably greater problem: affecting the Earth (system) as a whole, not a geographically limited part of it. Even so, the point remains; the notion that our human doings may have the potential to upset nonhuman nature so that nature will strike back at us, punishing our lack of modesty and prudence as the inferior party, is one of the oldest in recorded cultural history. Contrary to the portrayal of the shaman as the embodiment of superstition in "mythical" worldviews, falling short of what enlightened moderns deem a rational and scientific understanding of nature, the shaman may well have represented what in the end will prove the superior type of rationality.

VI

I mentioned the importance of James Lovelock's book *The Revenge of Gaia* to Latour's undertaking. Lovelock writes, "We have to understand that the 'silent spring' [Rachel Carson's 1962 book] did not come simply from poisoning by pesticides; the birds died because there was no longer space for them in our intensively farmed world. There are so many humans now aiming for a first-world lifestyle that we are displacing our partners on the planet, the other forms of life" (Lovelock 2007: 139).

Lovelock is better than Latour at describing not only the kind of loss of which birds are but one among numerous examples, but also why it is that natural creatures and processes long conceived as inert, passive, and indifferent are nothing of the sort, calling the lie to the said notion that, qua acted upon, has now brought about the quaking of the Earth that Serres drew attention to. "Unfortunately", Lovelock writes, "we are a species with schizoid tendencies, and like an old lady who has to share her house with a growing and destructive group of teenagers, Gaia grows angry, and if they do not mend their ways she will evict them" (Lovelock 2007: 60), there being, however, no place for them – us – to go to (leaving the one-percent "solution" – say, Mars – aside; see Hamilton 2013). According to the "Gaia hypothesis", that Lovelock and the American biologist Lynn Margulis first postulated in the early 1970s, the biosphere is an active, adaptive control system able to maintain the Earth in homeostasis. What has come to be called the Earth System is such as to behave "as a single, self-regulating system comprised of physical, biological and human components", thus breaking with the conventional wisdom according to which organisms adapt to, but do not change, their environments (2007: 32). The Earth is not something inanimate; it is the "largest living thing in the solar system", so that "unless we see the Earth as a planet that behaves as if it were alive, at least to the extent of regulating its climate and chemistry, we will lack the will to change our way of life and to understand that we have made it our greatest enemy" (2007: 21f.). We have, probably beyond a point of no return, since having set in motion positive feedback loops, upset the goal of the Earth's self-regulation, the goal being "to keep the Earth habitable for whatever are its inhabitants" (2007: 208).

What is most striking about Lovelock's approach as compared with Latour's is brought out in a passage like this:

> In several ways we are unintentionally at war with Gaia, and to survive with our civilization intact we urgently need to make a just peace with Gaia while we are strong enough to negotiate and not a defeated, broken rabble on the way to extinction. Can the present-day democracies, with their noisy media and special-interest lobbies, act fast enough for an effective defense against Gaia? We may need restrictions, rationing and the call to service that were familiar in wartime and in addition suffer for a while a loss of freedom. We will need a small permanent group of strategists who, as in wartime, will try to out-think our Earthly enemy and be ready for the surprises bound to come.
> (Lovelock 2007: 196f.)

That Lovelock, the intellectual father of the notion of Gaia and the person who, more than anyone else, helped inspire today's Earth system science, should resort to such warlike rhetoric, may be surprising; perhaps it has something to do with having experienced the war. At any rate, in Latour, born a generation later than Lovelock, such rhetoric is completely absent.

If we look in vain for Lovelock's analogy with war in Latour's account, what do we find in its stead?

"The religious origin of the ecological crisis is indisputable", argues Latour, in line with his fondness for invoking the work of Voegelin mentioned above. In the grand narrative proposed, "the Christians", "believing themselves to be attached to the Spirit, have lost the Earth. Believing that they are defending religion, they have driven everyone to assault the Earth through negligence" (Latour 2017: 210). However, "the fate of Christians is [. . .] of little importance compared to the loss of meaning imposed on materiality by the move to force it to become matter. It is here, really, where the major injustice lies, and this is what ultimately explains the Modern's insensitivity to what they do" (2017: 211). Latour concludes: "The Moderns are irreligious only in this: they neglect materiality" (ibid.). In saying this, and in adding that the contempt for matter goes back to Gnosticism, the influence from Serres is evident. Serres writes that "Modernity neglects, speaking in absolute terms. It cannot and will not think or act toward the global, whether temporal or spatial" (Serres 1995: 48).

Notice the ease with which Latour, like Serres before him, slides from the Christians to the Moderns, and from matter to materiality – the former two in the role of the biggest, most crude subject (agent) thinkable, the latter two in the role of object, of being acted upon. In a formulation like "Modernity neglects", who exactly is doing so, for what reasons and with what alternatives? If agency, to take Latour on his word, is always "shared" and never "autonomous", how can you account for these alleged facts in the case of "Modernity neglects"? The slide and the level of abstraction, calling for the capital "M", see to it that concrete entities – say, particular human agents or institutions – do not appear. The "grid" Latour talks about, and himself employs, is evidently too crude to allow for them, permitting only that everybody is depicted in like manner and exemplify the same broad categories.

If anything, to talk about negligence in trying to explain how it has come about that we in the modern Western world are causing the Earth – Lovelock's Gaia – to be in peril, amounts to a gross understatement. Negligence being much too mild, what would be a more accurate explanation?

Aggression would. But it is nowhere mentioned in Latour's book. That *aggression* is a deep-seated, if primitive and often contra-productive, affect in response to the experience of *being dependent on another being*, on its willingness to give as a gift what one's existence and well-being depend on, is an insight that goes back at least to Freud, one further explored in the work of psychoanalyst Melanie Klein (1988, 1989) in particular. The subjugation, torment, and all-out destruction, including killing, be it silent (systemic, abstract) or industrialized as in state-of-the-art factory chicken production, that all sorts of nonhuman animals are being subjected to and that is part of the crisis scenario that Lovelock addresses in his warlike language, are never identified as such by Latour, let alone reflected on with regard to what they tell us about our proneness to destroy the very beings and life-forms our rational intellect and our scientists tell us we depend upon for our very existence. Clearly the *modus operandi* at work in the ongoing "sixth extinction" in the history of planet Earth, distinct from all previous ones in the way human agents have their footprints all over the crime scene, is revealing something much more potent – destructively potent – than the bleak indifference that we looked at above – that's why Elizabeth Kolbert, in her Pulitzer-winning book on the subject, chose *An Unnatural History* as her subtitle (Kolbert 2014).

Melanie Klein's insight is the following: Being at the receiving end of the relationship, being vulnerable and at the mercy of life-sustaining sources that we cannot control, that are beyond our power and in that sense superior to us, strikes us as something difficult to acknowledge and to cope with; it triggers anxiety. This goes for the infant who senses his dependence on his mother and her willingness to give freely of her love and care, of her goodness, being a one-way affair and exposing the weakness and inferiority he has to endure, and hopefully one day accept, as someone ineluctably tied to her and needing her to exist. Writ large, it also – or so I suggest we expand on Klein's thesis – holds for humanity's relationship to Earth, our home, our *sine qua non*. The shock of dependence, of having to realize its reality, its power, amounts to a narcissistic blow – a blow that is not, however, "as such" wholly negative. That it need not be perceived as negative is demonstrated in cosmologies where humans' dependence on the nonhuman natural world is a constant ground for awe, humility, and gratitude toward that which is mightier and greater than we are, as we saw in the examples given in Chapter 4. By contrast, our Western, now rapidly globalizing anthropocentric culture (be it Christian or secular) is eminently ill-equipped to let us tolerate, in a positive sense, this dependence, making aggression, defiance, and denial in the face of what appears a thoroughly unwelcome reality the predominant and culturally induced response. Indeed, the more individualist the mentality fostered, and the more any reminder of dependence, of being indebted to something outside ourselves and our reach and control, is sensed as intolerable in that it threatens the self-sufficiency, independence, and autonomy that freedom, Western societies' trump value, is seen as craving, the more of an uphill struggle will be efforts to recognize the irremovable part played by dependence and the limits that it poses in any human individual's, any culture's life.

I am not suggesting that the importance of dependence-related and anxiety-driven aggression makes for a refutation of Latour's account and the factors he chooses to emphasize, following the lead of Voegelin and Serres. What I do suggest is that the absence of the role of aggression in that account is conspicuous, and that the burden of proof delegated to negligence is one for which it is unsuited, being much too mild and conflict-averse to do justice to the human-made destruction of the natural world; a destruction that does not grind to a halt despite overwhelming knowledge about its catastrophic consequences.

A related cost of Latour's silence on the role of aggression is his failure to address grief and mourning. How can you not dwell on them when writing a book about facing Gaia? A characteristic of the "depressive" position that Klein distinguishes from the primitive "paranoid-schizoid" one, mourning presupposes the ability to realize that aggression, the urge to destroy, and to destroy precisely what is good and worthy, *because* it is so and hence conducive to life, is not something that (only) *others* may experience and enact, but something of which I myself is undeniably capable. We humans have, through the centuries and the generations, caused such enormous amount of pain, inflicted such enormous wounds on the creatures with whom we co-inhabit this planet, as to be overwhelmed by grief and regret at the mere thought of it – if only we allow ourselves to think it, feel it. Only then, that is, may we start allowing for feelings of guilt over our part in destroying the "good object" (Klein), in this instance, planet Earth, and thereupon engage in efforts of reparation, seeking to restore what has been damaged, to affirm what is good and life-sustaining instead of envying it and wanting to spoil it insofar as anything pointing out how dependent and inferior is deemed intolerable.

Admittedly, making this point on the highest level of generality – "we humans" – loses out on all sorts of real-life differences and inequalities: the richest people in the arrangement of global capitalism are immeasurably greater destroyers and exploiters of the natural world than are the poorest (Kempf 2008), reflecting the ever-persistent pivotal role played by class division, a role ignored in much environmental philosophy, undermining its potency in terms of system critique.

Contradicting Latour's excitement over the proliferation of agencies, shared or not, he claims to see everywhere around him, especially when being with all sorts of scientists in the lab, this is George Monbiot in a *Guardian* column characteristically entitled "Our blindness to the living world is lethal, as we normalize the erosion of our environment, and the devastating losses to fragile ecosystems it brings":

> And the erosion continues. Few people younger than me know that it was once normal to see fields white with mushrooms, or rivers black with eels at the autumn equinox, or that every patch of nettles was once reamed by caterpillars. I can picture a moment at which the birds stop singing, and people

> wake up and make breakfast and go to work without noticing that anything has changed.
>
> (Monbiot 2018: 48)

Indifference may reign, as Monbiot has it: that is what the phenomenon of "shifting baseline syndrome" amounts to, namely that each new generation will take what is fact a degraded condition of nature as the nondegraded condition – as the normal experience (see Peterson 2009: 92). Unknowingly, each generation adjusts to a process of deterioration in its very life conditions that it had better not adjust to – adjustment showing itself not as the evolutionary asset that it is normally claimed to be, a prerequisite for thriving and surviving, but as their very opposite: as jeopardizing our ability to notice danger before it's too late.

But the eels, caterpillars, and birds will take notice, as do the penguins that van Dooren write about (referred to in Chapter 4); those directly affected by a phenomenon like loss of habitat-cum-home tend to do so. As van Dooren's colleague Deborah Bird Rose explains, the culturally reinforced "belief that the world is expressively inert does *not* destroy the world's expressivity. Modernity may damage our ability to hear, even as our violence [aggression – AJV] damages the world's ability to communicate, and we become caught up in an amplifying process of reduction: the refusal to hear stifles or destroys living things, thereby reducing the multivocal expressivity of the world". Nevertheless, Rose maintains, it is "essential that we do not mistake an inability to hear for an absence of communication, and not mistake loss for total death" (Rose 2004: 183).

As for the human-centered indifference that Monbiot focuses on, it is a reaction-formation, as the psychoanalyst would hasten to point out: not the real or original thing, not what set in motion the changes that Monbiot refers to, so as to lament them, grieve their impact, and admit the pain and guilt that follow, as Latour does not; indifference being, rather, a shield that we erect to protect ourselves from the consequences of our doings and, more deeply still, from the fact that they *are* of our doing, and so reflect on our capacities for harm, for doing wrong against what are so many sources of, magnificent manifestations of, life on Earth.

Against this background, is Lovelock justified in speaking of the "revenge" of Gaia? Latour does not question his doing so, yet (as we just saw) shies away from employing the belligerent language for which Lovelock has such a strong predilection. Isabelle Stengers, for one, does question it:

> To speak of the "revenge" of Gaia [. . .] is to mobilize a type of psychology that doesn't seem relevant. One doesn't struggle against Gaia. Even speaking of combating global warming is inappropriate. If it is a matter of struggling, it is against what provoked Gaia, not against her response. [. . .] The brutality of the intrusion of Gaia corresponds to the brutality of what provoked her, that of a development that is blind to its consequences, or which, more precisely, only takes its consequences into account from the point of view of the new sources of profit they can bring about. [. . .] Capitalism's mode of transcendence is not implacable, just radically irresponsible, incapable of answering for anything. [. . .] When it is a matter of capitalism, individual psychology is completely irrelevant. Capitalism must be understood instead as a mode of functioning, a machine, which fabricates its own necessity, its own actors, in every conjecture, and destroys those who haven't been able to saddle up for the new opportunities. [. . .] It is in the nature of capitalism to exploit new opportunities: *it cannot help doing it*.
>
> (Stengers 2015: 46, 52, 53, 54)

That readers of Lovelock will be unprepared for the close connection that Stengers suggests between Gaia's predicament on the one hand and the inner logic of capitalism on the other, is one thing. That readers of Latour's *Facing Gaia* are hardly any better prepared is another, more significant at that. Why?

Indeed: Why is Latour so silent on what has provoked Gaia, on the brutality that Stengers highlights? Why does Latour prefer to give such a pride of place to Modernity, Christianity, and religion in his historical account of the developments leading to the Anthropocene, to Gaia's revenge? To abstract and faceless entities instead of mundane, earthly and historical agents, such as specific humans, institutions, and corporations, for example? Are the agents, arrangements, and institutions involved really such as to call for what is Latour's take on the *history of ideas*? The resulting "idealism" and level of abstraction in Latour's account does not sit well with his stated commitment to a new materialism. But then again, the epistemological and ontological premises of the latter have not been sufficiently clarified and thought through by its proponents.

I have touched on these questions above, for example in my critique of Barad's reliance on posthumanist authors such as Donna Haraway. Whether posthumanist approaches are well-suited to help us comprehend what is at stake in the Anthropocene is of course big topic in its own right, one that I cannot do justice to here – but see Andreas Malm's 2018 book *The Progress of this Storm*, a book that only came to my attention after the completion of my manuscript, sharing many of my criticisms of posthumanism in general and of Latour in particular, though differing from the present work in focusing very one-sidedly on global warming as distinguished from on the natural world as explored in the ethnographic studies I have concentrated on.

Enough has been said in previous chapters to indicate that I am strongly inclined to answer the question raised in the negative: posthumanist approaches start out in a promising fashion by critiquing anthropocentrism in theory and in practice, yet falls into the one pitfall after the other when it comes to building a plausible alternative, be it analytically, be it substantively in terms of ontology, owing – in my view – in particular to the various proponents' commitment to constructionism. To be sure, a critic of posthumanism such as the Swedish anthropologist Alf Hornborg (2017) exaggerates when he states that an author like Haraway "aims at maximum unintelligibility and inaccessibility" in her ruminations about "tentacular thinking". But I concur – as would Stengers, I assume – when he says that "the promotion of post-humanist discourse is ultimately tantamount to looking away while neoliberal capitalism continues to destroy the planet". The case of Haraway aside, I have argued that in Latour's work there is a tendency to favor lofty abstractions and sweeping generalizations over analysis of particular agents, situations, and cases. The laudable and assumedly progressive and emancipatory goal of including as agents all – or nearly all? – beings, creatures, and entities that have for so long been excluded, rendered passive not active, objects not subjects, ends up offering a universe of actors/actants where the rhetoric of "all included, none excluded" carries the cost of depicting everybody in the same manner, as if making distinctions were tantamount to symbolic violence, to offending those being distinguished. Hornborg is right in suggesting that "the most problematic implication of Latour's ambition to dissolve the subject-object distinction is arguably not the fetishistic attribution of agency to nonliving entities, but the withdrawal of responsibility and accountability from human subjects". What Hornborg essentially questions, then, is the dissolution of the analytical boundary between the social and the natural, as well as of "the ontological distinction between sentient subjects and nonsentient objects" (Hornborg 2017).

What we witness, then, is the coming together of strange bedfellows, highlighting what amounts to an elective affinity between "post-nature", posthumanist constructivism à la Latour on the one hand, and the capital-backed techno-optimist fantasies of the prophets of geoengineering on the other. We first came across this type of affinity when discussing Barad's analysis of the brittlestar in Chapter 2. Christophe Bonneuil and Jean-Baptiste Fressoz capture the phenomenon well:

> In our late modernity, the "invisibilizing" of the limits of Earth is no longer just a result of an externalization (as a great outside that humans can draw from and jettison into without problem), but on the contrary of a radical internalization. This internalization is expressed in the efforts to measure ecosystemic functions in terms of financial flows, making a nature that is liquid and capitalizable even in its most intimate processes. Internalization into the market is backed by the ontological dissolution of nature of constructivist philosophies that deny its alterity in relation to humans, and from the geoconstructivist engineering project of all aspects of the Earth system, from the genome to the biosphere. [. . .] Fusion and omnipotence, these sentiments of early infancy, lie at the basis of such "post-nature" discourse, participating in the dream of a total absorption of nature into the commercial technosphere of contemporary capitalism.
>
> (Bonneuil and Fressoz 2016: 86, 220)

To be sure, the posthumanist plea to take the monopoly of agency away from humans, to distribute agency throughout the natural world and, in a spirit of political correctness and (I suppose) genuinely felt solidarity and egalitarianism, to admit agency to all kinds of living and non-living entities, may strike many a reader as greatly appealing – the plea seems to have a lot going for it, precisely because its declared philosophical adversary appears so dangerous in its real-life consequences. And there is no denying that, at least prima facie, there seem to be many commonalities between the panpsychism and animism that I have presented above and the various posthumanist attempts to explode and go beyond agency-anthropocentric-style so as to greatly, perhaps even limitlessly, expand agency's distribution outside of the narrow human-centered viewpoints, embodiments, and locations. And yet, what have we seen? We have seen that exposing the work by authors such as Latour and Bennett to the insights found in anthropologists studying nonmodern, non-anthropocentric cosmologies, in the form of so many specific cultural practices, *Lebenswelten* and life-forms, all of them organized around how to interact with those particular other-than-human others, at that time and place, allows us to recognize that what promised to be gained by exploding anthropocentrism and its peculiar way of contrasting culture and nature, all too easily end up effacing the very distinctions, the real-life presences of "othernesses", that any valid cosmological alternative has to prove itself able to do justice to, conceptually and practically. In a nutshell, the eagerness to remove humans from their (our) positing ourselves on a pedestal, as superior over all other beings in every important regard, ends up neglecting what *is* peculiar about us as one species among many. And failing to recognize our own peculiarities is no less serious a mistake than failing to acknowledge those of (our, many) others.

Drawing on the Australian philosopher Clive Hamilton, I now turn to a less theoretical, more political and, in that sense, more concrete look at what is at stake here.

VII

In his recent book *Defiant Earth*, Clive Hamilton argues that when the monopoly of agency is taken away from humans, the meaning of "agency" is changed so that it "no longer means to act with a purpose (as living things do)". Echoing my criticisms of Barad in Chapter 3, Hamilton holds that "collapsing intentionality into an ontological flat-world of influences leaves us bereft of the categories needed to understand power and politics". He refers, as I did above, to Hornborg's work to redeem agency by drawing distinctions between "non-living objects that have *consequences*, living entities other than humans that have *purposes* (due to their sentience and communication), and humans who have *intentions* (because they can reflect on their purposes). When agency is stripped of all element of choice it becomes mere influence, so we might summarize the distinctions as no choice, choice, and the capacity to make considered choices" (Hamilton 2017: 100). Specifically, Hornborg (2017) warns against dissolving the crucial difference between purposive agency and merely having consequences, since (as Hamilton observes) "in a world in which intentionality has been dissolved into networks or assemblages, there is no power, no freedom, and no morality" (Hamilton 2017: 100).

Hamilton's second point is about bad timing, and it is worth quoting in full:

> The post-humanist program of taking agency from the human monopoly and allocating it to processes or competing entities took place exactly at the time that in practice humans were accumulating and centralizing so much agency that we now rival the influence of the great forces of nature. In the Anthropocene, the lumpiness of the actual distribution of agency has become *more* pronounced just as the post-humanists are telling us that the world is flat. Only in the last two or three decades has the pre-eminence of human agency truly confronted us. No other force, living or dead, is capable of influencing the course of the Earth System *and* has the capacity to decide to do otherwise. Now *that* is agency. It is what makes humans *of* nature but also the freak of nature.
>
> (Hamilton 2017: 100f.)

The point is well taken, and it recalls my objections to Latour. It echoes Kate Soper's observation in her seminal study from 1995, *What is Nature?*, that, deliberately or not, "all ecological injunctions [. . .] are clearly rooted in the idea of human distinctiveness"; for "insofar as the appeal is to humanity to alter its ways, it presupposes our possession of capacities by which we are singled out from other living creatures and inorganic matter", responsibility being of course one crucial such capacity (Soper 1995: 40). But when Hamilton, conventionally at that, puts his emphasis on the capacities for agency that make humans unique and notions such as choice and responsibility meaningful, a disagreement emerges with Stengers' view, cited above, that the agency we should first of all focus on, because all-important in the Anthropocene, is that of capitalism, for which "individual psychology is completely irrelevant"; capitalism functioning like a machine "which fabricates its own necessity, its own actors" (Stengers 2015: 52). The question this view invites, and which remains unanswered in Stengers, is where rendering capitalism as a machine-like system respecting only the laws of its inner logic, that of supply and demand, leaves the prospect for *individual* human agents seeking to make a difference, that is to say, the sort of agents redeemed, if somewhat by default, through Hamilton's rejection of posthumanism. The closest that Stengers comes to pinpointing a place and a task for such "old-fashioned" and pre-Anthropocene individual agents, is in her discussion of refusal: the refusal to put oneself in the position of "the [EU] commissioners" and the CEOs, protesting against them and the machine-like system they represent; protesting, that is, that "it is without any mandate that they have defined the limits of political action by reference to their necessary subjection to what they call the laws of the market" (Stengers 2015: 127).

Although she makes no use of the distinction in play, I believe that Stengers' message here is none other than the one coming out of the Enlightenment: to be a human and moral agent is to refuse heteronomy and serfdom; in the very act of refusing them, of refusing compliance and the hand that one has been given (be it by the system, any system), the autonomy that is the hallmark of such agency is manifested – not as a lofty philosophical idea, but as a practice of which only such agents are capable, a practice to which they are entitled, whatever the circumstances. Indeed, this view seems to be the only one on which resistance can be built, the only alternative, that is, to seeing capitalism as a vast machine "which fabricates its own necessity, its own actors", not least since *that* way of describing the situation would only confirm (not challenge) the carriers of business-as-usual and TINA (there is no alternative), hence leaving everything the way it presently is.

And yet, the situation we – and the rest of those sharing the planet with us as their one and only home – find ourselves in now that the Anthropocene is here, and here to stay, is unprecedented. Its being so is what prompts Hamilton to contend that "against all ethics from Kant onwards, morality is not to be found in the realm of freedom but is rooted in the realm of necessity because our duty to care for the Earth must precede all others. It belongs to us alone" (Hamilton 2017: 53). This, of course, is the plea made all those years ago in Hans Jonas' *The Imperative of Responsibility*.

Interestingly, in light of my discussion in Chapter 4, Hamilton takes up Philippe Descola's *Beyond Nature and Culture*, where a central question is which unique feature can be held to separate humans from (the rest of) nature. He interprets Descola to the effect that we moderns would do better to go to premodern ontologies to understand the world around us rather than rely on what Descola refers to as "the tiny quantum by which we distinguish ourselves" from other objects, giving Hamilton the opportunity to make his point once again: "Well, that tiny quantum was enough to shift the Earth's geological arc and to do so more or less consciously" (Hamilton 2017: 103).

But this is not what Descola is saying. In the passage cited by Hamilton, he writes: "The ethnographical and historical ground that we have covered so far shows clearly enough that a consciousness of certain discontinuities between humans and nonhumans is not in itself enough to create a dualist ontology" (Descola 2013: 86). I take it that Hamilton agrees. Descola continues: "The multiplicity of forms of existence that we witness all around us may offer a more fertile terrain for ontological discriminations than the tiny quantum by which we distinguish ourselves from what Merleau-Ponty calls 'associated bodies'. The world presents itself to us as a proliferating continuum, and one would have to adhere to a truly myopic realism of essences to consider it cut up in advance into discontinuous domains that the brain is designed, always and everywhere, to identify in the same manner" (ibid.). That the world presents itself to us humans as a proliferating continuum is also something I would expect Hamilton to agree with. And if it stops doing so, it will be a truly remarkable development, a novelty for which humanity's collective role in "shifting the Earth's geological arc" is mainly responsible – this being a claim that Descola, for his part, would hardly disagree with.

If this is not the point on which Hamilton and Descola disagree, what is it? As we shall see, the real disagreement between them is relevant not only in light of my use of ethnographical studies in Chaper 4, but also because of the impasse – philosophically, morally, and politically – in which posthumanist discourse, including Latour's contribution, has landed us.

Even though Hamilton accepts that Western cosmology is only one among several ways of being in the world, as richly documented in Descola's work, his view is that "notwithstanding all their merits, pre-modern ontologies cannot help us now" (2017: 104). Hamilton argues like this:

> At the risk of speaking on their behalf, most Indigenous people understand that old worlds cannot be preserved except by rearticulating them in a dialogue with the modern world. Social scientists who call for a return to non-dualistic premodern ways of being – Descola even suggests we might find dead ones on library shelves and make them "come to life once more" – propose a political strategy that Indigenous people themselves typically do not entertain. [. . .] Indigenous people do not have the solution to the Anthropocene. [. . .] To turn to them for answers shoulders them with an impossible burden. We made the mess and "going native" ontologically is no answer.
>
> (Hamilton 2017: 105, 106)

But nowhere does Descola call for a return to "non-dualistic premodern ways of being". Rather, as he explicitly declares, his ambition is to show that a rejection of dualism – between culture and nature, and between humans and nonhumans – "leads neither to absolute relativism nor to a return to modes of thought that today's context has rendered obsolete", the task of anthropology being "to situate our own exoticism as one particular case within a general grammar of cosmologies" (2013: 85, 88). In this spirit, Descola concludes *Beyond Nature and Culture* by asserting that the purpose of anthropological theory is "not to propose models of communal life, of new forms of attachment to beings and things, or a reform of practices, mores, and institutions", adding, however, that such a reform is "indispensable", citing (without using the term "the Anthropocene") "the alarming degradation, as a result of human action, of the major bases of equilibrium in the biosphere". Even though the need for a reform is urgent, it would be "mistaken to think that the Indians of Amazonia, the Australian Aboriginals, or the Monks of Tibet can bring us a deeper wisdom for the present time than the shaky naturalism of late modernity" (2013: 405).

When I quote Descola's statements so extensively here, it is not only to show the inaccuracy of Hamilton's criticism. The points that Descola makes are important in their own right. If Hamilton's claim that "pre-modern ontologies cannot help us now" (Hamilton 2017: 104) nevertheless sounds correct, it has more to do with the thoroughness of modernity's destruction of the peoples which used to exhibit, say, an animist cosmology than with that cosmology's merits or lack of such – knowing full well that the merits are only worth their salt insofar as they both help sustain, and are themselves sustained by, the wider natural world that the cosmology-practicing culture (people) in question co-inhabit with so many "others". If the sort of cosmology (Descola) or ontology (Hamilton) that today rules on all continents has created a situation where no radical alternatives to it appear possible, this is so because *one* way of being in the world and of distinguishing between the different entities found there has come to dominate to the exclusion of all others: it is kind of tyranny, of tolerating no genuine "otherness" from itself at all, not only in a philosophical sense but in an exceedingly concrete one.

"There is irony", writes Deborah Bird Rose, "in the fact that Aboriginal people are being pressed to make transformations into modernity at a time when many of us are skeptical of modernity's capacity in the long run to sustain life on earth" (Rose 2004: 151), "irony" being of course an understatement. What we are talking about is not rival cosmologies in some abstract meaning but the way in which a given cosmology, being *that* group's such, at a specific time and place, is viable in terms of power, of ability to survive into the future given the resources it would take. The idea that the Indians of Amazonia or the Aboriginals in Australia should somehow come to "our" rescue and help us solve "our" problems is not only ludicrous; it is beside the point. It is a solution that no one has seriously proposed, least of all present-day anthropologists of the kind I have invoked, and as such a strawman. There is indeed something ill-conceived about the notion – or hope, or expectation – that the very peoples who have throughout the centuries been actively and aggressively "othered", meaning marginalized and ridiculed, should play a part in helping out their cultural oppressors in their hour of self-inflicted crisis – be it in the largely symbolic form of cosmological inspiration. The people in question, called indigenous by "us", have a hard enough time trying – fighting – to maintain what little is left intact of their cosmology and their land and the practices that have helped sustain both through the centuries. Not only are these people incapable of helping "us" in the mess that we have maneuvered ourselves into; they are scarcely in a position to fend for themselves, as their territory is taken over (often violently and illegally) by the local and regional proxies of global capital, creating in many such groups a conflict between the elders who want to preserve their traditional way of life, largely outside the market economy, and the youngsters who feel the attractions of urban lifestyle, modern technology, and the freedom of choice.

For lack of such in the books of Latour and his critic Hamilton alike, let me bring just one example:

> Saira Ka'apor used to patrol one of the world's most murderous frontiers, a remote and largely lawless region of the Brazilian Amazon where his indigenous community has fought for generations to protect their forest land. [. . .] That vigilante role came to an end last April when Saira was stabbed to death in Betel, a logging town close to Ka'por territory. [. . .] Numerous studies have shown protection of indigenous land is the most effective way to cut deforestation, but the Ka'por – like many other tribes – feel the police often work against them. According to Global Witness, Brazil is the deadliest country for environmental and land defenders, with 44 deaths recorded in 2017. Maranhao – the poorest state in the country – is among the worst affected. There were more death threats and attacks on indigenous groups here than anywhere else in 2016. [. . .] Almost certainly, Saira's killing will

> go unsolved. The nearest police station has yet [nine months after the murder] to open a case. Outside the indigenous community, nobody cares.
>
> (Watts 2018: 7)

One of the paradoxes of the anthropocentric mentality that is part and parcel of the Anthropocene should be mentioned here: for all "our" human-centeredness, we moderns – adults and children alike – are presently much more alarmed by the threat against species affected by biodiversity loss than by the extinction of life-forms and cosmologies that make up the loss of cultural diversity, the two types of losses being not separate but parallel, two sides of the same trend, hailed as "development", as pointed out in Chapter 4. The contrast is indeed a striking one, yet seldom reflected on: whereas the endangered mountain gorilla has become iconic, striking our children as "cute" and precarious (despite the animal being majestic – witness the similar case of the Arctic polar bear, nowadays sparking pity rather than fear, seeking in vain for food and for solid ice under its feet) and ringing the alarm bell whenever their dwindling numbers are mentioned, the people who co-exist with these animals and whose life-sustaining practices are unthinkable without them, attract none of the same sympathy and fascination, thus failing to mobilize the outside support needed to protect them from cultural extinction. A starved-to-death polar bear – and who hasn't seen the photos and videos by now? – is incomparably more concrete, more tangible and real, than a culture, the carrier of a particular cosmology, on the brink of extinction.

Of course, Hamilton does not literally mean that what anthropologists like Descola advocate is that indigenous people come to our rescue. Rather, what he has in mind is the idea of "our" "going native", by reaching out to "other people's worlds" so as to correct our ways, before it's too late. Hamilton puts his warning like this: "Looking upon Indigenous cultures with awe and regarding them as having magical potency is to fetishize them, a tendency now taken so far by some as to attribute to them the power to fix the climate and reverse the geological destabilization of the planet" (2017: 106). No cases to document this tendency are supplied, no names mentioned. To be sure, Hamilton does well not to attribute so naive a view to Descola, the one anthropologist whose work he discusses. It is hard to avoid the impression that Hamilton isn't really giving the voices of non-modern cosmologies a hearing, insisting – instead – that "the *only way forward* is to begin from where we are, in modernity, and from there work toward a 'beyond-modernity' way of being, a fifth ontology to add to Descola's four" (ibid.).

Nevertheless Hamilton does credit "certain Indigenous ontologies" with "something that ought to be recovered in a new Anthropocene way of being beyond modernity", namely their "cosmological sensibility", suggesting that "the absence of separation of the traditional Indigenous self from its natural world may hold a powerful message for how to live in the Anthropocene" (ibid.; see Clifford 2013). But then again, the *presence* of that very separation in modernity's thought and practice may have been held on to so long (not only individually so, but structurally and institutionally, making up the very fabric of our type of society) that the recovery of an alternative cosmological sensibility appears unrealistic, no less naive than the attribution of magical potency that we just saw Hamilton denounce. He does not say so, but it is as if the warning against the very course that modernity has taken – the one leading directly to our present worldwide ecological crisis – would need to have come from somewhere else than inside modernity itself, in order to qualify as *a sufficiently strong warning-cum-correction*, presupposing in its turn a sufficiently *different* way of seeing and practicing the relationship between humans and nonhumans from that perfected within modernity, with little or no tolerance for the value of anything that deviates from it. If this sounds abstract and theoretical, recall my earlier example about my son's successful shift from an animist participatory perspective on – being-in, communicating with – the natural world to a "rational" and objectifying one.

Though the question is ill-conceived, let me for the sake of argument put it like this: What is lacking in Hamilton's reflections about what our modern cosmology might learn from nonmodern ones such as explored by Descola (not to mention Ingold and Kohn, not discussed by Hamilton) is the element of dialogue, of genuine exchange between them, the one type of cosmology and the other. This would presuppose a considerable degree of reciprocity, the possibility of a genuine give-and-take between the different cosmologies in question and thus a high degree of symmetry. What is intellectually absent in Hamilton's approach on this score mirrors what has been and still is lacking in practice: that the potency of a so-called nonmodern "cosmological" sensibility is allowed to be more than purely ideational; that it is backed up by the extra-intellectual resources (economic, political) needed to make itself heard; in short, to establish itself as worth listening to, rather than being suppressed. Short of the power required to attain such a position, the alternative cosmologies that just might have it in them, philosophically, to serve as a corrective to a modernity closed in on itself and running amok have been utterly and effectively silenced. The bottom line for members of modernity that have abandoned their "childish" animism and adopted, as requested by everybody around them, a thoroughly objectifying stance where everything to do with intelligence, purposefulness, and value, indeed with taking up a point of view in and on the world, is considered exclusively human, is that "we" – our modern lifeform and the technologies that help improve it, globalize it – represent progress and development, and that whatever the casualties in terms of the twin-losses of biological and cultural (cosmological) diversity, *it is worth it*. If anyone dares suggest that we modern Westerners have something of true importance to learn from indigenous peoples not yet extinct, forget it. Truth be told, there is something annoying, something a-pain-in-the-ass-like about their very existence in today's world, against all odds. As long as they, or remnants of them, still exist, they are reminders that *the road taken was not inevitable and without alternatives*, alternatives abundantly existing, even thriving, at the historical start of the process commenced; reminders that things could be otherwise, that a radically different relationship between humans and the rest of nature is possible. The truly challenging view that nonmodern cosmologies represent, should they have the power required to do so loudly and clearly, namely that the costs brought by civilization are in many respects too destructive to be worth it, is felt to be intolerable and so will be knocked down immediately.

The interconnection and mutual dependence between biodiversity and cultural (including cosmological) diversity that I am stressing can be put positively as well as negatively. Put *positively*: to save the tiger, you must save the jungle, because the real entity is tigers-in-a-jungle, animals in their habitat, the whole (system) required for the individual exemplar to thrive and survive (Rolston 1988). And to save the jungle, or the forest, the people who live there must be allowed to continue doing so, based on their respect for all the kinds of creatures that go to make up the particular forest that is their common home, the lived biothic community theorized by philosophers such as Callicott, discussed in Chapter 4. Put *negatively*, the interconnection is that of violence twice over: the destruction of the natural world that is such a crucial aspect of the Anthropocene is inseparable from the *in situ* destruction of the peoples and cultures perceived as inferior, primitive, and worthless, as so many obstacles to the forward march of Reason, Progress, and Civilization. Historically, there was no dialogue between the proponents of the latter and the "savages" of the former. There was instead the radical asymmetry of power that facilitated the near-genocidal massacres of indigenous peoples around the world. Today, if not before, we realize just how efficient and rapid is the destruction of what remains of the natural world, of wildlife, of the habitats of iconic and not-so-iconic animals when the people who co-exist(ed) with them in a given region and territory are no longer present in large enough numbers and with the requisite resources to freely decide to do so – to live where they have lived for times immemorial.

In case someone had forgotten what this amounts to for those affected, consider this passage in Tzvetan Todorov's *The Conquest of America*:

> Enslavement reduces the other to the status of an object, which is especially manifested in the conduct [of the Spaniards] that treats the Indians as less than men: their flesh is used to feed the surviving Indians or even the dogs; they are killed in order to be boiled down for grease, supposed to cure the wounds of the Spaniards; thereby they are identified with animals for the slaughterhouse; all their extremities are cut off, nose, hands, breasts, tongue, sexual organs, thereby transforming them into shapeless trunks, as one might trim a tree; it is suggested that their blood be used to irrigate the fields, as if it were the water of a stream.
>
> (Todorov 1984: 175)

The all-important part played by violence both in the struggle against Indians and other native peoples and cultures, and in the ruthless exploitation of their land, cleared for the making of commodities and profits for the "superior" forces of civilization, is no less underplayed in Hamilton than in Latour, speaking instead, as both prefer to do, about "neglect". What ultimately led to the Earth's precarious situation in the Anthropocene, according to Hamilton, "was a power struggle between contending social forces, the forces of neglect – power-hunger, greed, growth fetishism, hedonism, and psychological weaknesses – against the forces of care: self-restraint, respect for the natural world, love of one's children, and the desire for civilization to flourish" (2017: 124; cf. 134, 135).

To suggest, contra Hamilton, that what paved the way for the present-day destruction of what remains of the natural world was the physical killing of what the anthropologist Eric Wolf calls "the people without history" (Wolf 1982), is of course to advance a simplistic and all but sophisticated thesis. It is to make a claim every bit as sweeping and generalizing as those that I criticized in the case of Latour. When I have raised it here, however, it is to correct the tendency that I see in philosophers (be they as different as Nagel and Hamilton) to discuss cosmology and ontology without taking into consideration the ways in which they are not ideational, abstract constructions that can be examined *sub specie aeternitatis*, assessing their purely intellectual credentials. Instead, they are *lived*, practiced, and acted upon as a given people's distinct way of being in the world, the natural as well as the cultural one, being two sides of the same reality.

Bibliography

Adorno, W. Theodor, 1973. *Negative Dialectics*. New York: Seabury Press.

Angus, Ian, 2016. *Facing the Anthropocene: Fossil Capitalism and the Crisis of The Earth System*. New York: Monthly Review Press.

Arendt, Hannah, 1978. *The Life of the Mind*. London: Secker & Warburg.

Barad, Karen, 2007. *Meeting the Universe Halfway: Quantum Physics and the Entanglement of Matter and Meaning*. Durham: Duke University Press.

Barth, Fredrik, 1987. *Cosmologies in the Making*. Cambridge: Cambridge University Press.

Basile, Pierfrancisco, 2009. "Back to Whitehead? Galen Strawson and the rediscovery of panpsychism", in Skrbina, D. (ed.), *Mind that Abides: Panpsychism in the New Millennium*: 179–199. Amsterdam: John Benjamins Publishing Company.

Bastian, Michelle, 2017. "Encountering Leatherbacks in Multispecies Knots of Time", in Deborah, B. Rose, et al., (eds.), *Extinction Studies: Stories of Time, Death and Generations*. New York: Columbia University Press.

Bateson, Gregory, 1972. *Steps to an Ecology of Mind*. New York: Chandler.

———, 2002. *Mind and Nature: A Necessary Unity*. Cresskill, NJ: Hampton Press.

Bennett, Jane, 2010. *Vibrant Matter: A Political Ecology of Things*. Durham: Duke University Press.

Berkes, Fikret, 2008. *Sacred Ecology*. London: Routledge.

Bhaskar, Roy, 2008 [1975]. *A Realist Theory of Science*. London: Routledge.

Blasdel, Alex, 2017. "The Anthropocene Age of Reckoning." *The Guardian Weekly*, July 14: 26–30.

Bohr, Niels, 1958. *Atomic Physiscs and Human Knowledge*. New York: Wiley.

Bonneuil, Christophe, and Jean-Baptiste Fressoz, 2016. *The Shock of the Anthropocene*. London: Verso.

Brennan, Teresa, 2004. *The Transmission of Affect*. Ithaca: Cornell University Press.

Butler, Judith, 1993. *Bodies that Matter*. New York: Routledge.

Butler, Judith, 1997. *Excitable Speech: A Politics of the Performative*. New York: Routledge.
———, *Precarious Life: The Powers of Mourning and Violence*. New York: Verso.
Callicott, J. Baird, 1989. *In Defence of the Land Ethic*. New York: SUNY.
———, 1994. *Earth's Insights*. Berkeley, CA: University of California Press.
———, 1999. *Beyond the Land Ethic*. New York: SUNY.
Carrington, Damian, 2017. "Sixth Mass Extinction is Here." *The Guardian Weekly*, July 21: 10.
Carson, Rachel, 1962. *Silent Spring*. Boston: Houghton Mifflin.
Ceballos, Gerardo, Paul Ehrlich, and Rodolfo Dirzo, 2017. "Biological annihilation via the ongoing sixth mass extinction signaled by vertebrate population losses and declines." *Proceedings of the National Academy of Sciences of the United States of America* 114 (30): E6089–E6096.
Clifford, James, 2013. *Returns: Becoming Indigenous in the Twenty-First Century*. Cambridge, MA: Harvard University Press.
Coleman, Sam, 2009. "Mind Under Matter", in David, Skrbina, (ed.): 83–108.
Crutzen, Paul, 2002. "Geology of Mankind." *Nature* 415: 23.
Deacon, W. Terrence, 2012. *Incomplete Nature: How Mind Emerged from Matter*. New York: Norton.
Descartes, Rene, 1978a [1637]. *A Discourse on Method*. London: Dent.
———, 1985 [1641]. *Meditations*. Cambridge: Cambridge University Press.
———, 1978b [1644]. *Principles of Philosophy*. London: Dent.
Descola, Philippe, 1996. "Constructing Natures: Symbolic Ecology and Social Practice", in P. Descola and G. Palsson, (eds.), *Nature and Society*: 82–102. London: Routledge.
———, 2013. *Beyond Nature and Culture*. Chicago: The University of Chicago Press.
Despret, Viviane, 2017. "Afterword: It is an entire world that has disappeared", in Deborah, B. Rose, et al., (eds.): 217–222.
Dodds, Joseph, 2011. *Psychoanalysis and Ecology at the Edge of Chaos*. London: Routledge.
Dreyfus, Hubert, 1991. *Being-in-the-World*. Cambridge, MA: MIT Press.
Eddington, Arthur, 1928. *The Nature of the Physical World*. New York: Macmillan.
Evans, Brad, and Julian Reid, 2014. *Resilient Life: The Art of Living Dangerously*. Cambridge: Polity Press.
Evernden, Neil, 1985. *The Natural Alien: Humankind and Environment*. Toronto: University of Toronto Press.
Fisher, Andy, 2013. *Radical Ecopsychology*. New York: SUNY.
Gadamer, Hans-Georg, 1975 [1960]. *Truth and Method*. London: Bloomsbury.
Geertz, Clifford, 1973. *The Interpretation of Cultures*. New York: Basic Books.
Georg, W. F. Hegel, 1969 [1812–14]. *Science of Logic*. Atlantic Highlands, NJ: Humanities Press.

———, 1975. *Aesthetics: Lectures on Fine Art*. Oxford: Clarendon.

———, 1977 [1807]. *Phenomenology of Spirit*. Oxford: Oxford University Press.

———, 1991 [1821]. *Elements of the Philosophy of Right*. Cambridge: Cambridge University Press.

Griffin, David R., 2007. *Unsnarling the World-Knot: Consciousness, Freedom, and the Mind-Body Problem*. Eugene, OR: Wipf & Stock.

Habermas, Jürgen, 1984 [1987]. *The Theory of Communicative Action, Vol.* I & II. Boston: Beacon Press.

Hacking, Ian, 1983. *Representing and Intervening*. Cambridge: Cambridge University Press.

Hamilton, Clive, 2013. *Earthmasters: The Dawn of the Age of Climate Engineering*. New Haven: Yale University Press.

———, 2017. *Defiant Earth: The Fate of Humans in the Anthropocene*. Cambridge: Polity Press.

Hannay, Alastair, 2018. Personal communication.

Haraway, Donna, 1997. *Modest Witness, Second Millennium: Feminism and Technoscience*. New York: Routledge.

Heidegger, Martin, 1962 [1927]. *Being and Time*. Oxford: Blackwell.

———, 1977 [1953]. *The Question Concerning Technology*. New York: Harper.

Hobbes, Thomas, 1966a [1651]. *Leviathan*, in *English Works*, *Vol. 3*. Aalen: Scientia.

———, 1966b [1655]. *De Corpore*, in *English Works*, *Vol. 11*. Aalen: Scientia.

Horkheimer, Max, and Theodor W. Adorno, 1979 [1947]. *Dialectic of Enlightenment*. London: Verso.

Hornborg, Alf, 2017. "Dithering While the Planet Burns: Anthropological Approaches to the Anthropocene." *Reviews in Anthropology* 46, 2–3: 61–77.

Hume, David, 1969 [1740]. *A Treatise of Human Nature*. Harmondsworth: Penguin.

Hylland, E. Thomas, 2006. *Engaging Anthropology: The Case for a Public Presence*. Oxford: Berg.

———, 2016. *Overheating: An Anthropology of Accelerated Change*. London: Pluto.

———, 2018. *Boomtown: Runaway Globalisation on the Queensland Coast*. London: Pluto.

Ingold, Tim, 2011a. *The Perception of the Environment*. London: Routledge.

———, 2011b. *Being Alive*. London: Routledge.

Jonas, Hans, 1984 [1979]. *The Imperative of Responsibility*. Chicago: University of Chicago Press.

———, 2001 [1966]. *The Phenomenon of Life: Toward a Philosophical Biology*. Evanston: Northwestern University Press.

Kant, Immanuel, 1979 [1781]. *Critique of Pure Reason*. London: Dent.

———, 1956 [1788]. *Critique of Practical Reason*. New York Macmillan.

Kempf, Herve, 2008. *How the Rich are Destroying the Earth*. Vermont: Chelsea Green.

Klein, Melanie, 1988. *Love, Guilt and Reparation and Other Works 1921–1945*. London: Virago.

———, 1989. *Envy and Gratitude and Other Works 1946–1963*. London: Virago.

Klein, Naomi, 2014. *This Changes Everything: Capitalism vs the Climate*. London: Allen Lane.

Kohn, Eduardo, 2013. *How Forests Think: Toward an Anthropology Beyond the Human*. Berkeley, CA: University of California Press.

Kolbert, Elizabeth, 2014. *The Sixth Extinction: An Unnatural History*. London: Bloomsbury.

Kovel, Joel, 2002. *The Enemy of Nature*. London: Zed Books.

Kuhn, Thomas, 1970 [1962]. *The Structure of Scientific Revolutions*. Chicago: University of Chicago Press.

Latour, Bruno, 2004. *Politics of Nature: How to Bring the Sciences into Democracy*. Cambridge, MA: Harvard University Press.

———, 2017. *Facing Gaia: Eight Lectures on the New Climate Regime*. Cambridge: Polity Press.

Lee Mueller, Martin, 2017. *Being Salmon, Being Human*. Vermont: Chelsea Green.

Leopold, Aldo, 1949. *A Sand County Almanac*. Oxford: Oxford University Press.

Levinas, Emmanuel, 1985. *Ethics and Infinity*. Pittsburgh: Duquesne University Press.

Levinas, Emmanuel, 1990. *Difficult Freedom. Essays on Judaism*. Baltimore: Johns Hopkins University Press.

———, 1991. *Totality and Infinity*. Dordrecht: Kluwer Academic Publishers.

———, 1992. *Otherwise than Being or Beyond Essence*. Dordrecht: Kluwer Academic Publishers.

Lovelock, James, 2007. *The Revenge of Gaia*. Harmondsworth: Penguin.

MacIntyre, Alasdair, 1981. *After Virtue*. London: Duckworth.

Magdoff, Fred, and John Bellamy Foster, 2011. *What Every Environmentalist Needs To Know about Capitalism*. New York: Monthly Press.

Malm, Andreas, 2018. *The Progress of This Storm. Nature and Society in a Warming World*. London: Verso.

Mathews, Freya, 1991. *The Ecological Self*. London: Routledge.

McGinn, Colin, 1991. *The Problem of Consciousness*. Oxford: Basil Blackwell.

Merchant, Carolyn, 1990 [1980]. *The Death of Nature*. New York: Harper.

———, 2016. *Autonomous Nature*. New York: Routledge.

Merleau-Ponty, Maurice, 1962 [1945]. *Phenomenology of Perception*. London: Routledge and Kegan Paul.

Midgley, Mary, 2010. *The Solitary Self: Darwin and the Selfish Gene*. Durham: Acumen.

———, 2014. *Are You an Illusion?* Durham: Acumen.

Monbiot, George, 2018. "Our Blindness to the Living World is Lethal." *The Guardian Weekly*, January 5: 48.

Moore, Jason, 2015. *Capitalism in the Web of Life*. London: Verso.

Morton, Timothy, 2013. *Hyperobjects. Philosophy and Ecology After the End of The World*. Minneapolis: University of Minnesota Press.

———, 2016. *Dark Ecology: For a Logic of Future Coexistence*. New York: Columbia University Press.

———, 2018. *Being Ecological*. London: Pelican.

Nagel, Thomas, 1979. *Mortal Questions*. Cambridge: Cambridge University Press.

———, 1986. *The View from Nowhere*. Oxford: Oxford University Press.

———, 2002. *Concealment and Exposure*. Oxford: Oxford University Press.

———, 2012. *Mind and Cosmos: Why the Materialist Neo-Darwinian Conception of Nature is Almost Certainly False*. Oxford University Press.

Næss, Arne, 1989. *Ecology, Community, and Lifestyle*. Cambridge: Cambridge University Press.

Oreskes, Naomi, and Erik M. Conway, 2014. *The Collapse of Western Civilization: A View from the Future*. New York: Columbia University Press.

Peirce, Charles Sanders, 1998. *The Essential Peirce: Selected Philosophical Writings, Vol. 2*. Bloomington: Indiana University Press.

Peterson, Anna L., 2009. *Everyday Ethics and Social Change*. New York: Columbia University Press.

Rolston, Holmes, 1988. *Environmental Ethics: Duties to and Values in the Natural World*. Philadelphia: Temple University Press.

Rose, Deborah Bird, 2004. *Reports from a Wild Country: Ethics for Decolonisation*. Sydney: UNSW Press.

———, 2011. *Wild Dog Dreaming. Love and Extinction*. London and Charlottesville: University of Virginia Press.

———, et al, eds., 2017. *Extinction Studies: Stories of Time, Death, and Generations*. New York: Columbia University Press.

Rouse, Joseph, 1996. *Engaging Science*. Ithaca, NY: Cornell University Press.

Russell, Bertrand, and Alfred North Whitehead, 1913. *Principia Mathematica*. Cambridge: Cambridge University Press.

———, 1927. *The Analysis of Matter*. London: Routledge.

———, 1948. *Human Knowledge: Its Scope and Limits*. London: Routledge.

———, 1956. "Mind and Matter", in *Portraits from Memory*. Nottingham: Spokesman.

Sahlins, Marshall, 1976. *Culture and Practical Reason*. Chicago: The University of Chicago Press.

Sartre, Jean-Paul, 1956 [1943]. *Being and Nothingness*. New York: Washington Square Press.

Scanlon, Thomas, 1998. *What We Owe to Each Other*. Cambridge, MA: Harvard University Press.

Schütz, Alfred, 1970. *On Phenomenology and Social Relations*. Chicago: The University of Chicago Press.

Seager, William, 2006. "The "Intrinsic Nature" Argument for Panpsychism", in G. Strawson et al., (eds.): 127–145.

Searle, John, 1992. *The Rediscovery of the Mind*. Cambridge, MA: MIT Press.

Searles, Harold, 1972 [1960]. *The Nonhuman Environment in Normal Development and in Schizophrenia*. New York: International Universities Press.

Serres, Michel, 1995 [1990]. *The Natural Contract*. Ann Arbor: The University of Michigan Press.

Shapin, Steven, and Simon Schaffer, 1985. *Leiathan and the Air-Pump. Hobbes, Boyle, and the Experimental Life*. Princeton: Princeton University Press.

Shepard, Paul, 1982. *Nature and Madness*. Athens, GE: The University of Georgia Press.

Sherburne, Donald, 1981. *A Key to Whitehead's Process and Reality*. Chicago: The University of Chicago Press.

Simons, Peter, 2006. "The Seeds of Experience", in G. Strawson et al., (eds.): 146–150.

Skrbina, David, 2007. *Panpsychism in the West*. Cambridge, MA: MIT Press.

———, ed., 2009. *Mind that Abides: Panpsychism in the New Millennium*. Amsterdam: John Benjamins Publishing Company.

———, 2015. *The Metaphysics of Technology*. New York: Routledge.

Soper, Kate, 1995. *What is Nature?* Oxford: Blackwell.

Stengers, Isabelle, 2000. *The Invention of Modern Science*. Minneapolis: The University of Minnesota Press.

———, 2011. *Thinking with Whitehead*. Cambridge, MA: Harvard University Press.

———, 2015. *In Catastrophic Times: Resisting the Coming Barbarism*. Lüneburg: Open Humanities Press.

Strawson, Galen, 2006. "Realistic Monism: Why Physicalism Entails Panpsychism", in G. Strawson et al., (eds.): 3–31, *Consciousness and Its Place in Nature*. Exeter: Imprint Academic.

Taylor, Charles, 1985. *Human Agency and Language*. Cambridge: Cambridge University Press.

———, 1989. *Sources of the Self*. Cambridge: Cambridge University Press.

———, 1995. *Philosophical Arguments*. Cambridge, MA: Harvard University Press.

Taylor, Matthew, 2018. "Warnings of "Disastrous" Plastics Binge." *The Guardian Weekly*, January 5: 12.

Taylor, Paul, 1986. *Respect for Nature*. Princeton: Princeton University Press.

Todorov, Tzvetan, 1984. *The Conquest of America: The Question of the Other*. New York: Harper Perennial.

Turnbull, Colin, 1965. *Wayward Servants: The Two Worlds of the African Pygmies*. London: Eyre & Spottiswoode.

van Dooren, Thom, 2014. *Flight Ways: Life and Loss at the Edge of Extinction*. New York: Columbia University Press.

Vetlesen, A. Johan, 1994. *Perception, Empathy, and Judgment: An Inquiry into the Preconditions of Moral Performance*. University Park, PA: Penn State Press.

———, and Harald Jodalen, eds., 1997. *Closeness: An Ethics*. Oslo and Boston: Scandinavian University Press.

———, 2009. *A Philosophy of Pain*. London: Reaktion Books.

———, 2015. *The Denial of Nature: Environmental Philosophy in the Era of Global Capitalism*. Abingdon and New York: Routledge.

Watts, Jonathan, 2018. "Brazil Killing Splits Loggers and Villagers." *The Guardian Weekly*, January 26: 7.

Weber, Max, 1949 [1904]. *The Methodology of the Social Sciences*. Glencoe, IL: The Free Press.

Weintrobe, Sally, ed., 2013. *Engaging with Climate Change: Psychoanalytic and Interdisciplinary Perspectives*. London: Routledge.

Whitehead, Alfred North, 1929. *The Function of Reason*. Princeton: Princeton University Press.

———, 2007 [1920]. *The Concept of Nature*. New York: Cosimo Classics.

———, 1967 [1925]. *Science and the Modern World*. New York: The Free Press.

———, 1978 [1929]. *Process and Reality: An Essay in Cosmology*. New York: The Free Press.

———, 1966 [1933]. *Adventures of Ideas*. New York: The Free Press.

———, 1968 [1968]. *Modes of Thought*. New York: The Free Press.

Williams, Bernard, 1978. *Descartes: The Project of Pure Enquiry*. Harmondsworth: Penguin.

Williams, Robert, 1997. *Hegel's Ethics of Recognition*. Berkeley: The University of California Press.

Wittgenstein, Ludwig, 1958. *Philosophical Investigations*. Oxford: Blackwell.

Wolf, Eric, 1982. *Europe and the People Without History*. Berkeley, CA: The University of California Press.

Index